INNOVATING IN A SECRET WORLD

INNOVATING IN A SECRET WORLD

The Future of National Security and Global Leadership

TINA P. SRIVASTAVA

Potomac Books
AN IMPRINT OF THE UNIVERSITY OF NEBRASKA PRESS

 Potomac Books is an imprint of the University of Nebraska Press.
Manufactured in the United States of America.

Library of Congress Cataloging-in-Publication Data
Names: Srivastava, Tina P., author.
Title: Innovating in a secret world: the future of national security and global leadership / Tina P. Srivastava.
Description: Lincoln: Potomac Books, an imprint of the University of Nebraska Press, 2019. | Includes bibliographical references and index.
Identifiers: LCCN 2018056227
ISBN 9781640120860 (hardback: alk. paper)
ISBN 9781640122086 (epub)
ISBN 9781640122093 (mobi)
ISBN 9781640122109 (pdf)
Subjects: LCSH: Technology and state—United States. | National security—Technological innovations—United States. | Intelligence service—Technological innovations—United States. | Public-private sector cooperation—United States. | BISAC: POLITICAL SCIENCE / Government / National. | POLITICAL SCIENCE / Political Freedom & Security / International Security.
Classification: LCC T21 .S75 2019 |
DDC 355/.033073—dc23 LC record available at https://lccn.loc.gov/2018056227

Set in Sabon Next LT Pro by E. Cuddy.

CONTENTS

List of Illustrations vii
List of Tables ix
Acknowledgments xi

Introduction: Maintaining a Technological Edge for National Security 1
1. The Emergence of Open Innovation 7
2. The World of Secret U.S. Government R&D 29
3. Success and Failure in Secret U.S. Government Technology Innovation 41
4. Practical Consequences and Perverse Incentives 55
5. Secrecy versus Open Innovation 79
6. Incentives for Innovation 103
7. The Path to Long-Term Improvement 133

Notes 143
Bibliography 155
Table of Authorities 177
Index 179

ILLUSTRATIONS

1. The role of technology innovation strategies 4
2. The "iron triangle" of systems engineering 9
3. Categorization of open technology innovation strategies 16
4. Example of employing design variables 18
5. Bubble chart showing frequency of design variable choices 20
6. Functional decomposition of the innovation/R&D process 22
7. Functional decomposition of VC-arm approach 27
8. Technology readiness levels 30
9. Stakeholder value network 32
10. CIA website careers page 34
11. CIA's first official tweet 35
12. Increase in total U.S. secrecy orders 39
13. New secrecy orders 40
14. FANG-1 open technology innovation approach 48
15. DARPA FANG tiered challenges 53
16. Declining U.S. share of satellite exports 57
17. M67854-01-C-0001 contract 70
18. Potential misalignment of needs and incentives 73
19. SBIR awards over time for one SBIR shop 76
20. ISCAP decisions on declassification 96
21. Appeals to ISCAP 97
22. Declassified document, "The Uraba Massacres" 98

23. Declassified document, "The President's Daily Brief" 99
24. Reasons for incomplete invention disclosure 105
25. Funding source for incomplete invention disclosures 106
26. Employee reports on invention disclosures 107
27. Apparatus for generating mechanical power 111
28. USS *Aylwin* 112
29. USS *Warrington* under construction at the Cramp Shipyard 113
30. Tension created by Authorization and Consent 116
31. Inventions developed at government vs. private expense 122

TABLES

1. Representative examples of open technology innovation in the U.S. commercial sector 13
2. Representative examples of open technology innovation strategies pursued by the U.S. government 24
3. Template company commercialization report 77
4. Demonstrated pattern of reversal of security classification policy 92

ACKNOWLEDGMENTS

Thank you to my friends in the military, government, industry, and academia for their valuable input, insights, and support over the course of this research.

I extend special thanks to Oli de Weck, PhD, Professor of Aeronautics and Astronautics and Engineering Systems at MIT, and chair of my doctoral committee at MIT. His guidance and support over the course of this research have been invaluable.

Also I want to thank John Akula, PhD, JD, Senior Lecturer in Law at the MIT Sloan School of Management, for his insight and thought-provoking discussions, especially with regard to the legal aspects of my research.

I am grateful to Brent Appleby, Deputy VP of Engineering for Science and Technology at Draper Laboratory, for his support, perspectives on innovating in closed environments, and feedback that improved the quality of my work.

My thanks go to Nathan Wiedenman, former DARPA Program Manager of Adaptive Vehicle Make (AVM), which included FANG, for his on-the-ground expertise and experience, which contributed greatly to my research.

I would like to thank Mike Sarcione, former Senior Principal Engineering Fellow at Raytheon. He is recognized by government and industry as a world expert on radar. It has been my privilege to work alongside him, and I value his continued mentorship.

Thank you also to Tom Swanson, Abby Stryker, and the whole Potomac team.

I am grateful to my loving husband and my family, whose unwavering support and encouragement made this possible.

Finally, you the reader should be thankful to Scott Cooper, my editor, without whom this book would be much longer and still a work in progress.

INNOVATING IN A SECRET WORLD

Introduction

Maintaining a Technological Edge for National Security

On May 2, 2011, U.S. Navy SEAL operators killed Osama bin Laden, founder of al-Qaeda and mastermind of the terrorist attacks of September 11, 2001. Headlines highlighted the cutting-edge technology that was critical to the success of their mission:

> Osama bin Laden's Death Reveals the Value of State-of-the-Art Technology
> —*Venture Beat*[1]

> Attack on Bin Laden Used Stealthy Helicopter That Had Been a Secret
> —*New York Times*[2]

> New Details Emerge On The Surveillance Technology Used To Hunt Osama Bin Laden
> —*Popular Science*[3]

The SEALS used, among other things, two stealth helicopters, infrared imaging equipment, instantaneous DNA analysis technology, and secure wireless communication channels with enough bandwidth to live-stream compressed video. Increasingly our national security depends on access to the most sophisticated and advanced technology—like that used in the bin Laden mission.

In the mid-1990s the Clinton administration described the importance of science and technology to national security when laying out America's national security strategy: "Our defense science and technology investment enables us to counter military threats and to overcome any advantages that adversaries may seek. It also expands the military options available to policymakers."[4] American leadership around the world is critical to peace and security. While this leadership takes many forms, such as economic, cultural, and scientific, the importance of military leadership cannot be ignored. Our allies also depend on American military strength.

That strength requires a foundation of technological superiority. The Clinton administration further recognized the importance of technological superiority: "*Technological superiority underpins our national military strategy*, allowing us to field the most potent military forces by making best use of our resources, both economic and human. *It is essential for the United States to maintain superiority in those technologies of critical importance to our security.*"[5] The Bush administration echoed this sentiment in 2005, recognizing that "*the preservation of technological superiority is a key component to our national security strategy*."[6] Technological superiority also contributes to national security because it helps prevent "strategic surprise," defined as "the sudden realization by an organization that it has operated on the basis of an erroneous threat assessment resulting in an inability to anticipate a serious threat to its vital interests."[7]

Our national security depends on advancements in science and technology.[8] Research and development investments are critical in "enabling us to stay at the cutting edge of new developments so that our Armed Forces remain the best trained, best equipped, and best prepared in the world."[9]

The basic premise behind this strategy is that R&D will bring forth technology innovation in the defense sector that in turn creates technological superiority that supports national security, a sequence of outcomes illustrated as follows:

R&D → Technology innovation → Technological superiority → National security

More recently, President George W. Bush and President Barack Obama also emphasized the importance of maintaining a technological edge to support national security and the interests of the American people. As the 2015 National Security Strategy states, "We will safeguard our science and technology base to keep our edge in the capabilities needed to prevail against any adversary. . . . The United States will use military force, unilaterally if necessary, when our enduring interests demand it: when our people are threatened; when our livelihoods are at stake; and when the security of our allies is in danger."[10]

After the fall of the Soviet Union, the United States became the world's sole hegemonic power. Dominance in the technological domain (e.g., precision weapons, stealth, and communications networks) have contributed significantly to this position, especially as related to military strength. Today, however, America's position of technological strength is being challenged in an increasingly multilateral world. Continued technological superiority depends on sustaining and improving a robust pipeline of technology innovation. As the Task Force on Innovation wrote in 2012, "The United States still leads the world in research and discovery, but *our advantage is rapidly eroding*, and our global competitors may soon overtake us."[11] The investments in R&D that countries, including China and India, are making "reflect an acknowledgment that *science and technology once helped make the U.S. the most powerful nation on earth*, and similar power could belong to the nation that most *successfully harnesses its intellectual resources and cultivates innovation* within its borders."[12]

It is no longer enough to support R&D and expect technology innovation, technological superiority, and national security to follow. For the United States to maintain its position requires being the foremost at *effectively* transitioning R&D investment into technology innovation.

Innovation strategy can be understood as the science of improving the yield from this process—that is, increasing the technology innovation that comes out of R&D efforts, as figure 1 shows. Over the years, we have figured out that innovation is a science. While new innovation can appear spontaneous—e.g., the result of serendipitous tinkering—it can also be cultivated and accelerated through purposeful strategy and rigorous efforts.

Public investment in R&D has been critical in creating the necessary environment to foster innovation.[13] Yet in the political and economic climates of the last decade, U.S. government budgets for R&D have declined. In the face of these cutbacks, technology innovation strategies must increase the return on public

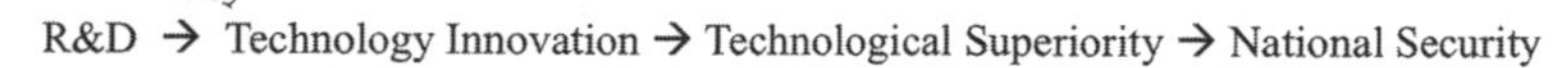

Fig. 1. The role of technology innovation strategies in the process from R&D to innovation supporting national security. Created by the author.

investment in R&D and be even *more* effective at yielding technology innovation.

Beyond financial constraints, innovating is harder today because the next generation of technology requires interfacing with systems that have thousands of technical interconnections crossing many disciplines.[14] Consider just one example from Ashton Carter, secretary of defense in the Obama administration. As Secretary Carter noted in his book, *Keeping the Edge: Managing Defense for the Future*, "The electronic warfare suite aboard tactical aircraft now under development is a complex system uniting radar, targeting, communications, electronic countermeasures, and attack warning functions previously attached to the aircraft system as separate subsystems."[15] The time it takes to deliver an initial operating capability (IOC) increases with complexity for aerospace systems. As Lt. Col. Nathan Wiedenman, a program manager of the Defense Advanced Research Projects Agency (DARPA), said, "A big part of the problem is that even though the systems we're building are much more complex than they have been in the past, the way we engineer defense systems hasn't fundamentally changed in a long time."[16]

Technology innovation strategies themselves must evolve if they are to succeed in the face of financial pressures and increasing technological complexity. The process of innovation itself is complex, difficult to model, and multifaceted. The United States needs further analysis of pioneering technology innovation strategies that have shown promise. The National Research Council, whose charter is to improve U.S. government decision making and public policy, noted in a 2012 report that, unfortunately, the tried-and-true technology innovation strategies that have been honed over decades with strong grounding in theory are unlikely to be adequately modernized to yield the demanding results necessary given today's dynamic environment. "A new policy approach is required," the council concluded, "one based on a richer understanding of the complexity and global dimensions of innovation ... [and] policies also need to take into account the increasingly global and open nature of the innovation process."[17]

The commercial sector has benefited greatly from new innovation strategies, such as open innovation and gamification, resulting in advancements in technology areas such as genomic mapping, artificial intelligence, and autonomous driving. In the defense sector, however, certain endemic constraints, such as a complex intellectual property and regulatory environment, have hindered the successful deployment of new innovation strategies. Also R&D targeted toward national security—referred to in this book as *secure U.S. government* R&D—is cloaked in secrecy, which is seemingly at odds with open innovation strategies. The innovation pipeline for national security is influenced and governed by a web of executive strategy, legal codes, agency regulations, and judicial checks and balances. Technology innovation strategies must be supported by a robust legal and regulatory framework lest the U.S. position in the world be threatened.

About This Book

> As the pillars of the U.S. innovation system erode through wavering financial and policy support, the rest of the world is racing to improve its capacity to generate new technologies and products.
>
> —National Research Council

It is imperative that secure U.S. government R&D leverage the most advantageous technology innovation strategies to counteract this trend and cultivate technology innovation in areas of importance to national security. This book is about how to make that happen.

I begin with a detailed examination of innovation and U.S. government landscapes, and then I examine current trends in secure U.S. government R&D. From there I evaluate the crucial question of whether technology innovation strategy execution is unintentionally leaving certain innovations behind or unintentionally precluding certain classes of innovators from participating. I identify unintended consequences or emergent behavior resulting from the complex dynamics of the system, including the legal framework in which technology innovation must exist. I conclude with a call to action for changes at the system level, including in the regulatory regimes that govern secure U.S. government R&D environments. Such changes are required to overcome the constraints endemic to these environments so that the United States can cultivate the open innovation strategies that can enable, accelerate, and enhance innovation in areas of importance to national security.

ONE

The Emergence of Open Innovation

The commercial sector is successfully leveraging innovation strategies to accelerate technology innovation and enhance business growth. Recent commercial examples of disruptive technology range from smart phones to cloud computing to blockchain to virtual and augmented reality. Given that the commercial sector is often where we find the state of the art in technology innovation, it makes sense to consider whether the innovation strategies employed in that sector can serve as a model for the U.S. government and secure R&D. Among the strategies employed by the commercial sector is an emerging class known as *open innovation* that touts breakthrough success in achieving technology innovation in terms of the time and cost required to innovate, as well as the diversity and novelty of ideas generated.

Open innovation is about broadening participation in innovation beyond an individual organization or division traditionally assigned to perform specific R&D activities. Henry Chesbrough, credited with coining the term, wrote in 2003, "Open innovation is a paradigm that assumes that firms can and should use external ideas as well as internal ideas, and internal and external paths to market, as the firms look to advance their technology."[1]

Chesbrough goes on to explain that an organization, to accelerate internal innovation, should leverage the "purposive inflow" of knowledge from outside the organization. Among the many approaches for establishing such a purposive inflow of knowledge to solve a problem or advance technology, three stand out:

1. Fostering competition
2. Providing a means to share ideas and collaborate

3. Offering incentives to innovators to participate

To understand how these work, consider three brief open technology innovation examples widely regarded as successful.

Three Open Innovation Examples

The first example is one of *fostering competition*. In 1996 the X Prize Foundation publicly offered $10 million to any nongovernmental organization able to "build and launch a spacecraft capable of carrying three people to 100 kilometers above the Earth's surface, twice within two weeks."[2] It was a lofty goal. (The prize was later renamed the Ansari X Prize, following a multimillion dollar donation to the cause by entrepreneurs Anousheh and Amir Ansari.) In 2004 Mojave Aerospace Ventures (MAV) won the prize. The MAV team, led by Microsoft co-founder Paul Allen and noted aerospace engineer Burt Rutan, successfully achieved the space flight goal. Perhaps even more significant was the fact that their win signaled a dramatic shift in the field of space flight. It opened the door to a new private space industry and commercial space flight.

This competition-centric open innovation strategy succeeded in achieving resource leveraging; the winning team's estimated development costs were $25 million, or 2.5 times the prize money. The strategy created reputational incentives beyond the monetary prize that contributed to the winning team's investing far more than the prize money to win. "Over the course of the competition, 26 teams invested over $100 million in aggregate for research and development in suborbital space flight."[3] Incredible advancements in technology areas such as propulsion and spacecraft reusability were made, not just by the winning team. That is, the entire field of space flight benefitted from the competition.

Systems engineering uses the "iron triangle" to depict the constraints on projects, as figure 2 shows. Open technology innovation challenges can be set up to constrain any of the triangle's dimensions. Some set a limit on the duration of the competition, and the winner is often the best solution achieved within that time frame. The Ansari X Prize had a fixed goal and scope but an unrestricted schedule.

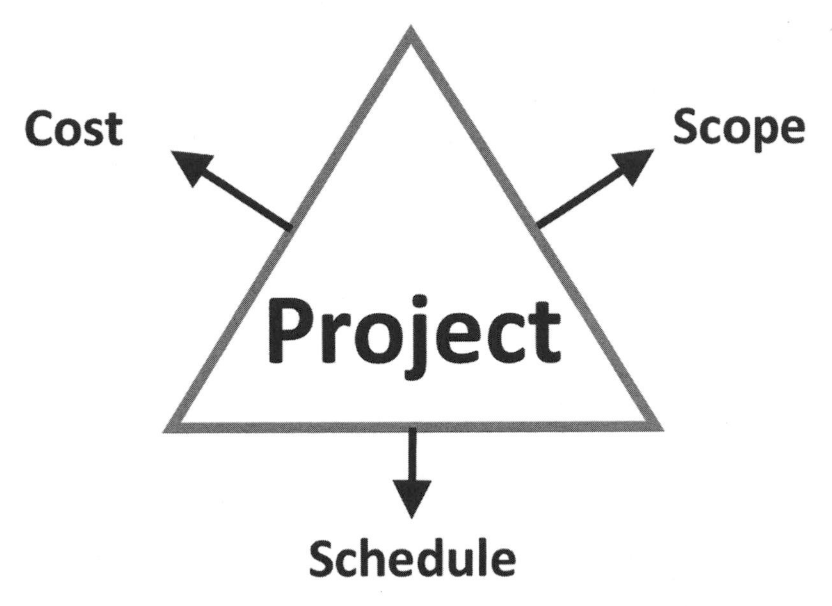

Fig. 2. The "iron triangle" of systems engineering. Source: de Weck, Lyneis, and Braha, "ESD.36 System Project Management Lecture 1."

The second example illustrates *providing a means to share ideas and collaborate*. The start-up company Quirky created a community of innovators and employs a unique intellectual property (IP) incentive structure that helps find the best ideas and then leverages that community to develop those ideas. Quirky has established an environment in which innovators have access to collective knowledge and the ability to share ideas to collaborate.

The Quirky approach involves a three-stage business model. First, the company asks its community to submit ideas. Second, inventors can ask members of the community for help in improving or developing their ideas in exchange for a share of the royalties. Finally, Quirky takes the top product ideas, builds those products, and sells them, sharing some of the revenues with the community according to the IP agreement.[4]

Quirky has also established a structure that incentivizes potential innovators to participate. The company offers royalties to idea contributors, with the goal of increasing access to potentially valuable inventions. The IP and reward policies are tied to the innovation strategy. The approach has paid off: Quirky has been successful in generating a large number of ideas from a diverse population in a short period of time, receiving about a thousand ideas per week.[5]

Like Quirky, the third example also offers *incentives to innovators to participate*. The idea is that the larger the pool of potential solvers to a problem, the more likely it is that a solution will be found. If we can incentivize more innovators to participate, drawing from an interdisciplinary pool of innovators bringing different skills and experiences, then we may be able to solve previously unsolved problems or decrease the time it takes to reach the solution.

Innocentive is an online platform for organizations to launch prize-based open challenges to solve technology problems. The pharmaceutical company Astra Zeneca, consulting firm Booz Allen Hamilton, and health-care provider Cleveland Clinic have all posted "challenge problems" at Innocentive for anyone in the community to solve.

The Innocentive community of "solvers"—individuals from the community who submit solutions—includes millions of people. The challenge success rate is 85 percent.[6] Innocentive promotes an interdisciplinary perspective because solvers are not necessarily experts in the field of the particular challenge topic.

The potential power of the Innocentive approach can be seen in the experience of Roche Diagnostics, a world leader in *in vitro* diagnostics, which posted a problem pertaining to product quality control. The company had been working on the problem for fifteen years; the Innocentive responses all came within sixty days. Notably the Innocentive community identified every approach Roche had already tried in addition to identifying a solution.[7]

The Roche example shows that the cost of open innovation can be significantly lower than the use of internal R&D alone. The cost issue is particularly relevant due to the risk factors associated with dedicating resources to an unsolved problem. As the Innocentive model shows, an open innovation strategy can be architected to have comparatively less risk than internal R&D, such as capping the time allocated to solving the problem and leveraging pay-for-performance such that the solution seeker pays only for the winning idea.[8]

All in all, Roche's experience with Innocentive was a success. The high quality of ideas—which included not only what Roche had generated internally, but also a winning solution—*enhanced* innovation; Innocentive *accelerated* innovation by generating a large number of ideas in a short time frame; and most important for Roche, this open approach helped solve a problem and thus *enabled* innovation.

Limitations in Certain Environments

Today there is no consensus on what defines open technology innovation strategy. Researchers have recognized this gap and, as one wrote, "Open innovation is a young concept that includes several meanings. . . . Researchers have different definitions and interpretations about open innovation."[9] There has never before been a taxonomy developed to help classify these various strategies so that they can be critically analyzed and their potential applications to different situations evaluated. As a result, many implementations of open technology innovation are "copycat" and often not grounded in academic theory.

The gap in open innovation research widens in the U.S. government context. There have been few studies of open innovation in the government and fewer still in the areas of national security and defense. The U.S. government remains "in the early stages of adoption of open innovation," and academia is still "in the process of understanding relevant issues."[10] Academic researchers of open innovation in the U.S. government have focused almost exclusively on the delivery of public services, not on technology innovation.[11]

Several endemic aspects of secure U.S. government R&D environments help explain the dearth of open innovation examples. Chesbrough identified three limitations to the application of open innovation that correspond directly to these aspects.[12] First, he explained that current open innovation strategies do not work well when proprietary or other restrictions prevent a solution seeker from being able to fully define the problem needing to be solved. Organizations seeking solutions to problems sometimes purposely obscure aspects of the problem to avoid tipping off their competitors, but this approach can hinder innovation. Chesbrough advised that providing a complete description of the problem is necessary to elicit helpful innovations. Unfortunately in U.S. government environments, security classification often prevents problems from being fully defined. The "competitors" in this context may be global adversaries.

Second, open technology innovation strategies do not work well when tacit knowledge is required to fully understand the problem. If the problem must be experienced firsthand and cannot be adequately described in words, it does not lend itself to be solved with open innovation.[13] Unfortunately this situation is also characteristic of secure U.S. government R&D.

Third, most open innovation strategies are best suited to point solutions—that is, solving one particular problem without regard to related issues—and do not work well in the context of technical interdependencies. However, as discussed above, the technology needed to advance national security is often complex, interconnected, and dynamic.[14] For certain technology areas critical to national security, innovators require a prerequisite knowledge base of the complex systems and technical interdependencies involved.

To overcome these limitations first requires a broader understanding of what open technology innovation would mean in a U.S. government context. A starting point is the taxonomy I mentioned above, and the remainder of this chapter introduces a taxonomy to further not only research, but also, and more important, the *implementation* of open innovation in secure U.S. government environments, where it is so desperately needed.

Building a Taxonomy

To begin let's explore the breadth of open technology innovation strategies that have been used in the U.S. commercial sector, affording us an opportunity to see trends and design variables that will drive the analysis of potential opportunities for open innovation in U.S. government programs. In this context the U.S.

commercial sector includes the industries and organizations that benefit civilian commercial interests either directly (business-to-consumer, or B2C) or indirectly (business-to-business, or B2B), including both for-profit and nonprofit organizations not owned and operated by the U.S. government and that may be privately or publicly held.

Table 1 shows a variety of open technology innovation strategies used in the commercial sector. The eighteen examples were selected from more than one thousand to illustrate the breadth of the solution space. These examples span industries from software to appliances to retail and others. In some cases these open innovation examples are entities themselves, such as Quirky, Kiva, and Threadless, and other examples are part of a larger company, such as Sandisk Ventures, Cisco Entrepreneur in Residence (EIR), and LEGO Ideas. The list also includes platforms that support multiple open innovation approaches, such as Innocentive.

TABLE 1. Representative examples of open technology innovation in the U.S. commercial sector

Ansari X Prize	Organization that launches public competitions to bring about radical breakthroughs, incentivized by multi-million dollar prizes. No fixed timeframe; goals are generally achieved in about eight years.
Quirky	Online platform for inventors to submit ideas that are developed and manufactured for sale. Inventors are rewarded with royalties. The time frame from idea to sale is about a year.
Innocentive	Crowdsourcing platform for organizations to seek ideas to solve specific technical problems. Innovators are financially incentivized; awards vary but are often \$10,000–\$80,000. Time frames are fixed, such as sixty days.

Google Ventures	Venture capital arm of Google that invests in start-ups. Investments are "non-strategic," meaning they are not limited to those synergistic with Google. Over fifty investments a year, usually a few million dollars each.
SanDisk Ventures	Strategic investment arm of SanDisk. Launched in 2012 with a $75 million fund to invest in companies synergistic with SanDisk.
Dell Technologies Capital	Strategic investment arm of Dell, including EMC, VMware, and other divisions. About twelve investments a year, usually a few million dollars each.
Indiegogo	Crowdfunding platform to raise funds for projects, charities, and start-ups. Can be a fixed duration (such as one month) or a fixed amount (such as $3,000). Indiegogo takes about a 4 percent fee. Incentives for funding a project range from pure charity to T-shirts to equity in a start-up.
Kickstarter	Crowdfunding platform, competitor to Indiegogo.
LEGO Ideas	Website for users to submit ideas that might be turned into LEGO sets for purchase. Innovators are incentivized through recognition, complimentary LEGO sets, and royalties, such as 1 percent of net sales.
Topcoder	Organization that administers programming contests of fixed duration, ranging from a few hours to weeks. Dozens of challenges are open at a time, each with a financial incentive, such as $500. Reputational incentives include a Topcoder Ranking.
Threadless	Company that creates and sells T-shirts based on designs submitted by users. Users receive a share of profits.

Zazzle	Online marketplace for users to create and sell arts-and-crafts-based products. Community includes designers, makers, and buyers.
GE: FirstBuild	"Micro-factory" for producing low quantities of innovative appliances; leverages 3D printing for rapid prototyping. Innovators are incentivized through recognition and royalties, such as 0.5 percent of sales.
Cisco Entrepreneur in Residence (EIR)	"Incubation program" for start-ups relevant to Cisco. Incubation period is limited to six months.
Linux / The Linux Foundation	Widely used open source computer operating system. Linux has spawned a community that has developed tools to manage contributions.
Kiva	Nonprofit that allows people to lend money via the internet to low-income/underserved entrepreneurs and students. Loan duration varies. Average loan is $400.
MIT Clean Energy Prize	Annual innovation contest for students. $100,000 grand prize, plus prizes targeted to specific categories, such as improving energy usage.
Foldit	Online puzzle game to discover native protein structures, leveraging gamification to encourage participants to fold proteins.

It should be noted that outcomes associated with many open innovation implementations tend to be documented by the initiators and sponsors themselves rather than in independent, objective assessments. The latter would be indicative of a higher level of maturity in the field of open innovation; the former illustrates that open innovation remains at an early stage.

Based on the list in table 1, the taxonomy of open innovation strategies in figure 3 categorizes those that have been pursued in

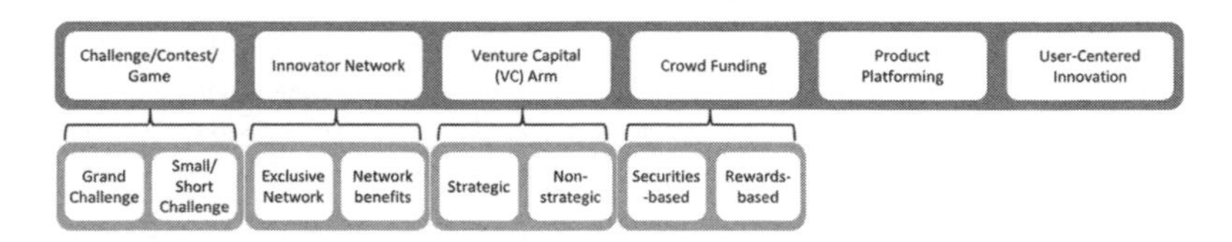

FIG. 3. Categorization of open technology innovation strategies in the U.S. commercial sector. Created by the author.

the U.S. commercial sector in recent decades. Let's look at these in a bit more detail. "Challenge/Contest/Game" and "Innovator Network" correspond to the examples listed above in this chapter. "Challenge/Contest/Game" includes "Grand Challenges" such as the Ansari X Prize, which set a lofty goal and offered prize money as a reward. "Innovator Network" leverages a community of innovators to solve technology challenges, such as in the Quirky and Innocentive examples. The innovators in the community often have varied backgrounds and are not necessarily subject-matter experts, a situation that can lead to interdisciplinary work and generation of ideas that cross knowledge boundaries.

"Venture Capital (VC) Arm" is a twist on traditional funding strategies. VC arms can be "self-directed new venture groups charged with moving the firm into a new market."[15] They often invest in start-ups with a technology focus synergistic with the firm's growth areas and that have potential for merger or acquisition at a later point.

"Crowdfunding," a relatively recent concept, involves posting ideas on a designated online portal and raising money by soliciting contributions from a large number of contributors. Accepting small contributions creates a lower threshold to participate. There are several varieties of crowdfunding. In microfinancing-based crowdfunding, the money contributed is in the form of a loan—as in the example of Kiva, a nonprofit founded in 2005 that connects potential lenders with borrowers in multiple categories. In rewards-based crowdfunding, such as through the increasingly well-known Indiegogo and Kickstarter platforms, the money can be a donation that is rewarded with items such as a T-shirt or even an early version of the product. Finally, in securities-based crowdfunding, money is contributed by what the U.S. Securities and Exchange Commission (SEC) calls "accredited investors" in exchange for equity. Recent SEC rules allow anyone, not just accredited investors, to invest an amount based on the annual income and net worth of the investor.[16]

"Product Platforming" involves opening a firm's base product platform for others to add on features and components, such as apps. Google's partially open source Android operating system is a commonly cited example. Innovations built on the platform are often incremental rather than "disruptive" because they conform to an existing platform. Many of this strategy's benefits tend to be in product adoption and network externalities rather than in the enhancement of innovation capability through a unique strategy or structure.

"User-Centered Innovation" seeks to benefit from loyal product users' experiences by involving them in the product development and redesign process. These users, sometimes called lead users, are often evangelists of a product and proactively customize products to their own needs. These customizations may highlight areas for new feature development. For example, a user of a commercially available snow blower may make innovative modifications to it for increased throughput, using odds and ends such as zip ties rather than professional design and manufacturing tools. Users might be willing to share their innovations with manufacturers for free because they would benefit from their modifications being incorporated professionally into future versions of the product.[17] Like "Product Platforming," this strategy tends to be limited to incremental innovations.[18]

Each strategy's implementation is defined by design variables—for example, whether a financial prize is offered or who retains intellectual property rights. It should be noted that it is possible to change design variables within an open innovation strategy without changing the overarching strategy itself. For example, a monetary prize and some IP protection could be offered in the context of a "Challenge/Contest/Game," as in the case of the Ansari X Prize, or there could be a non-monetary prize and no IP protection, but in both cases it would still be a Challenge/Contest/Game.

Design Variable	Options	VC Arm	Challenge/ Gamification	Innovator Network	Crowd Funding
Objective is to get more money invested in solving the problem than just what you have to spend	Yes No	Yes No	Yes No	Yes No	Yes No
Objective is to increase general/public awareness of the problem/need	Yes No	Yes No	Yes No	Yes No	Yes No
Objective is to lower the threshold for participation/contributing to solving problem	Yes No	Yes No	Yes No	Yes No	Yes No
Objective is to get user/customer buy-in early	Yes No	Yes No	Yes No	Yes No	Yes No
Recurrence Model	O – One-time T – Tiered P – Platform	O T P	O T P	O T P	O T P
IP protection offered	N – No M – Medium Y - Yes	N M Y	N M Y	N M Y	N M Y
Commitment/Investment required to participate (financial, duration of competition, # participants, Deliverable).	S – Small M – Medium L - Large	S M L	S M L	S M L	S M L
# of winners	Z – Zero F – One/Few M – Many N – N/A	Z F M N	Z F M N	Z F M N	Z F M N
Audience (# participants or # of competitors)	H – Hundreds T – Thousands M - Millions	H T M	H T M	H T M	H T M
Individual/team. Team size	I – Individual S – Small C - Consortium	I S C	I S C	I S C	I S C
Prize Offered	M – Monetary Prize N – Non-Monetary Prize X – No Prize	M N X	M N X	M N X	M N X
Prize distribution	F – First Place Only T – Top Few E – Everyone N – N/A	N	F T E	F T E	F T E
Participant	S – School I – Individual C - Company	S I C	S I C	S I C	S I C

Fig. 4. Example of employing design variables within an open technology innovation strategy. Created by the author.

Figure 4 illustrates how the choice of design variables affects the implementation of a strategy. Objectives are not typically considered design variables, but they are included here to aid in determining which strategy to pursue. "Recurrence model" refers to whether the approach is a one-time effort, recurring, or tiered (as in the example of FANG, discussed in chapter 3 below). The "Number of winners" matters because it may speak to the nature and complexity of the goal advertised by the sponsors. Is it a contest that is so difficult that no one wins? Is the concept of a winner even applicable? Considering the "Audience (number of participants or number of competitors)" as a design variable may mean modifying the implementation with respect to marketing efforts, platform capacity, resources allocated, or type of platform.

As for the other variables shown, there may be no "IP Protection Offered," full protection if the IP was developed independently ("yes"), or partial protection ("medium"), which is less than the full protection that would have been afforded had the idea been developed independently and which generally manifests in the form of a royalty-free, non-exclusive license for the sponsoring organization. This analysis uses this coarse granularity, as it is sufficient to understand the impact on instances of open technology innovation strategies in this context.[19]

The "Commitment/investment required to participate" is, in essence, a measure of the resource barrier to entry for an innovator. An approach could require anything from a paragraph write-up of an idea that takes an individual no more than a few hours to complete to a multi-year commitment by a well-funded team or the construction of a physical prototype. This speaks as well to "Team size": can an individual participate alone, or is a team needed? How large a team? Does it require a large consortium to collaborate as a single team? The decision may be left to the participants or may be stipulated by the sponsors.

Two variables are prize-specific. The "Prize offered" could be a monetary prize or nothing more than a T-shirt for participating. The variable may not capture reputational or intangible rewards. For example, simply being listed as a high scorer in Topcoder or Foldit may be a "prize" since such recognition is considered a high honor in certain communities. There may be multiple prizes or different prizes, depending on how a participant finishes—that is, in which "place" ("Prize distribution").

Finally, there may be requirements that stipulate whether or how the participation is limited—for instance, whether it is only for students or individuals generally or, whether, as in some cases, companies are eligible ("Participant").

As one can imagine, there are thousands of possible variations to each open innovation strategy given these design variables. In fact, for figure 4 those variations number 26,244 within each strategy.

In the "bubble chart" in figure 5 I apply the taxonomy to the

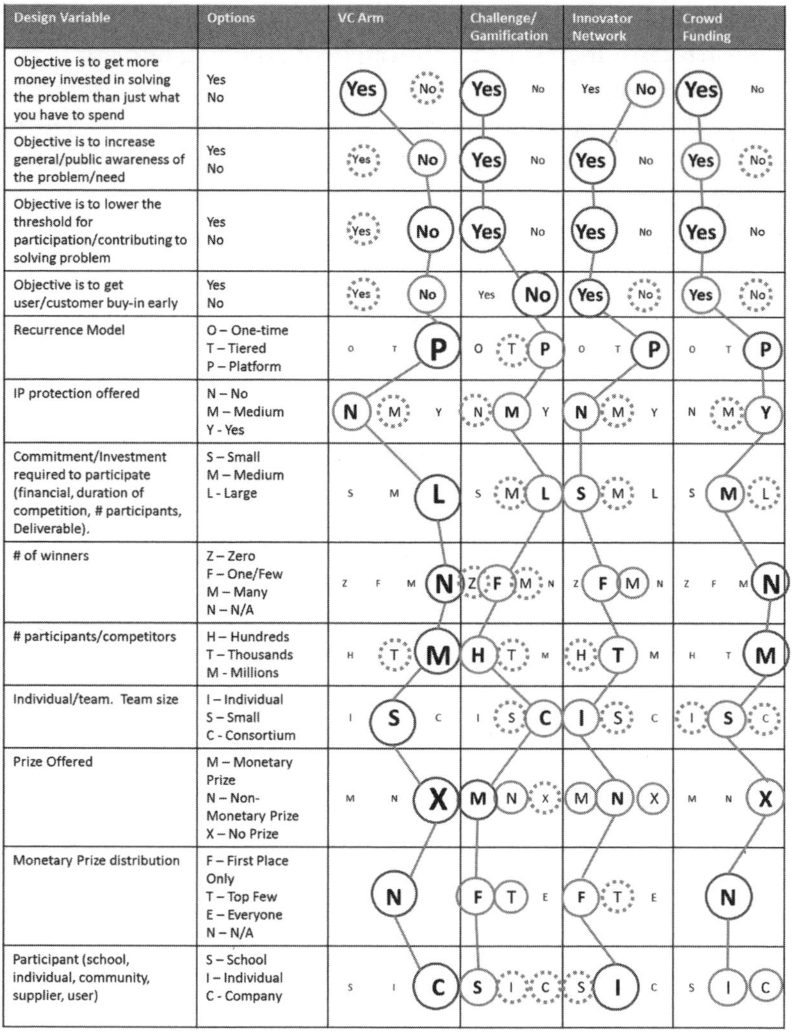

Design Variable	Options	VC Arm	Challenge/ Gamification	Innovator Network	Crowd Funding
Objective is to get more money invested in solving the problem than just what you have to spend	Yes No	Yes No	Yes No	Yes No	Yes No
Objective is to increase general/public awareness of the problem/need	Yes No	Yes No	Yes No	Yes No	Yes No
Objective is to lower the threshold for participation/contributing to solving problem	Yes No	Yes No	Yes No	Yes No	Yes No
Objective is to get user/customer buy-in early	Yes No	Yes No	Yes No	Yes No	Yes No
Recurrence Model	O – One-time T – Tiered P – Platform	O T P	O T P	O T P	O T P
IP protection offered	N – No M – Medium Y - Yes	N M Y	N M Y	N M Y	N M Y
Commitment/Investment required to participate (financial, duration of competition, # participants, Deliverable).	S – Small M – Medium L - Large	S M L	S M L	S M L	S M L
# of winners	Z – Zero F – One/Few M – Many N – N/A	Z F M N	Z F M N	Z F M N	Z F M N
# participants/competitors	H – Hundreds T – Thousands M - Millions	H T M	H T M	H T M	H T M
Individual/team. Team size	I – Individual S – Small C - Consortium	I S C	I S C	I S C	I S C
Prize Offered	M – Monetary Prize N – Non-Monetary Prize X – No Prize	M N X	M N X	M N X	M N X
Monetary Prize distribution	F – First Place Only T – Top Few E – Everyone N – N/A	N	F T E	F T E	N
Participant (school, individual, community, supplier, user)	S – School I – Individual C - Company	S I C	S I C	S I C	S I C

FIG. 5. Bubble chart showing which design variable choices are more frequently used than others. LEGEND: The larger the font size and the larger the circle size, the more frequent the use of an approach compared to others. Solid circles represent more frequent approaches than dashed circles. A small font without a circle represents a rare design choice. The zig-zag line going down through the morphological matrix in each column shows the most frequent approach. Created by the author.

examples shown in table 1. A large bold font and large solid circle represent design choices that are more frequently used than others. A small font without a circle represents where current implementations of this strategy have not made this design choice or where such a choice is rare; these are design choices that may not yet have been explored. For example, one design choice not yet explored is in the VC Arm strategy design variable for commitment/investment required by the participants. Equity investments come with vesting terms that require the start-up founders to stay committed to the company for a specified period of time, usually years. The taxonomy highlights that VC arms could consider an alternate approach where investments could be made in efforts that last a shorter amount of time and require less of a commitment by potential participants.

The framework of design variables in figure 5 does not distinguish one strategy from another, so the taxonomy has been further developed using the systems engineering method of functional decomposition. Functional decomposition is a tool used in systems engineering wherein a complex system or process is broken down into smaller parts. As applied here, first, the innovation process was decomposed into its functions: selecting the problem, generating the ideas, commercializing/productizing the innovation, financing the innovation, facilitating, and selecting the solution/idea to be financed or commercialized. Next, each function was attributed to the party that performs the function. Figure 6 applies this functional decomposition to the open innovation strategies, showing that the functional steps are the same across strategies and that the distinguishing characteristic among the strategies is the party that performs each functional step. The figure shows, as a baseline, the traditional closed internal R&D strategy in the U.S. commercial sector (bottom row).

The functional decomposition is also applied to traditional U.S. government contracting (second row from bottom of fig. 6)—specifically most military branches for large defense contractors, as well as small businesses through the Small Business Innovation Research (SBIR) program (as discussed in chapter 2 below). The government is generally responsible for problem selection.

Strategy	Instance	Problem Selection	Idea Generation	Commerc-ialization	Financing	Facilitator	Solution Selection
VC Arm	Google Ventures	Start-up	Start-up	Start-up	Google	Google Ventures	Google Ventures
	Sandisk Ventures	Start-up	Start-up	Start-up	Sandisk	Sandisk Ventures	Sandisk Ventures
	Dell Ventures	Start-up	Start-up	Start-up	Dell	Dell Ventures	Dell Ventures
	Cisco EIR	Proposer	Proposer	Proposer	Cisco	Cisco EIR	Cisco EIR
Challenge/ Gamification	Ansari X-Prize	X-Prize	Participant	N/A	Sponsor	X-Prize	X-Prize (sets rules)
	Topcoder algorithm competitions, tournaments	Topcoder	Participant	N/A	Sponsor	Topcoder	Topcoder (sets rules)
Innovator Network	Threadless	Threadless	Participant	Threadless	Customer	Threadless	Customer
	LEGO Ideas	LEGO	Participant	LEGO	Customer	LEGO	Customer
	Linux	Linux	Community	Linux	Community	Linux	Community
	Innocentive	Sponsor	Participant	Sponsor	Sponsor	Innocentive	Sponsor
	Topcoder Challenges	Sponsor	Participant	Sponsor	Sponsor	Topcoder	Sponsor
CrowdFunding	Kickstarter	Participant	Participant	Participant	Community	Kickstarter	Community
	Indiegogo	Participant	Participant	Participant	Community	Indiegogo	Community
	Kiva	Loanee	Loanee	Loanee	Investor	Kiva	Investor
	Quirky	Participant	Participant	Quirky	Community	Quirky	Community
	GE FirstBuild	Participant	Participant	GE	Customer	GE	Customer
	Zazzle	Participant	Participant	Zazzle	Customer	Zazzle	Customer
Traditional Gov't R&D Contracting		Gov't	Contractor	Contractor	Gov't	Acq. Agency	Gov't
Internal R&D		Company	Company	Company	Company	N/A	Company

FIG. 6. Functional decomposition of the innovation/R&D process and its application to open innovation strategies and traditional U.S. government contracting. LEGEND: Dotted boxes = party responsible for idea generation responsibility; horizontal lined boxes = party responsible for facilitation; shaded boxes = third-party participant. If that same party performs other responsibilities, the same formatting is used. Created by the author.

Contractors propose ideas, and their proposals must show evidence that they are able to commercialize the idea or technology successfully. The government selects and finances the project with the assistance of an acquisition agency.

It is in the context of functional decomposition that crowdfunding warrants a bit more explanation as an open innovation strategy. Unlike other strategies, it does not open up who is generating ideas but rather opens up other functional areas of the process—for example, modulating the investor sources rather than the innovator pool—to create more opportunities for innovation. In the end crowdfunding expands the innovation pipeline by bringing people into the process; that fact alone makes it an open innovation strategy. Had these additional people not been brought in, certain innovations may not have been realized and certain problems may have been left unsolved.

Each strategy can have openness at each functional step. Crowdfunding has openness at "financing the innovation" and, relatedly, at "selecting the solution/idea to be financed or commercialized," and it has the highest degree of openness at these steps because anyone can finance the innovation or select the solution/idea by financing it and making it possible to be commercialized.

Within crowdfunding two sub-strategies are differentiated according to the party performing the function of commercialization: in some cases the participant/loanee performs this function, whereas in others it is the facilitator. Perhaps a third sub-strategy still to be explored by existing crowdfunding platforms is to have an altogether different party perform the function of commercialization.

Degrees of Openness

It becomes apparent through this investigation that the term "open innovation" itself may be somewhat of a misnomer since many so-called "open" strategies can be and are applied in "closed" environments as well, such as through internal innovation contests limited to a company's employees. This suggests that open technology innovation strategies are really about the points and degrees of openness incorporated into their implementation.

With respect to the U.S. government, the traditional contracting process shown in figure 6 is arguably already more "open" than traditional U.S. commercial-sector R&D because participation by companies that are not government entities is invited through government contracting. However, government contracting has been locked into a single open innovation strategy for many years. The difference between a given open innovation strategy and traditional U.S. government contracting is the way in which the strategy innovates on the traditional contracting approach. Understanding this opens a pathway for traditional government contracting to adopt different open innovation strategies.

Government Forays into Open Technology Innovation

For the United States to maintain its global position of technological superiority, the government's technology innovation strategies must evolve. One way this evolution is occurring is through experimentation with open innovation, although this is in the early stages of adoption as the U.S. government initiates varying levels of open innovation with varying degrees of success. Table 2 is a representative list of examples of open technology innovation pursued by the U.S. government in recent years. It includes both platforms and instances that leverage platforms.[20]

TABLE 2. Representative examples of open technology innovation strategies pursued by the U.S. government

NeedipeDIA (the platform)	Public and classified versions of a website where Defense Intelligence Agency (DIA) end users post latest mission needs.
NeedipeDIA need: Prevent Strategic Surprise	A posted need on NeedipeDIA seeking methods and tools to build upon its warning apparatus.
Challenge.gov (the platform)	Platform for the U.S. government to post challenges. Postings are open for roughly three months with rewards up to $500,000.

DARPA Grand Challenge/ DARPA Urban Challenge	A series of challenges starting in 2004 with a $2 million prize to develop driverless cars that can complete an off-road and urban course.
DARPA Robotics Challenge	Building off of DARPA Urban Challenge, this instance seeks to spur development of semi-autonomous robots.
FANG	Three-tiered challenge to test DARPA tools and construct an amphibious vehicle; first and only U.S. government open innovation program seeking to tackle classified requirements and data.
DARPA Shredder Challenge	One-time challenge in 2011 to reconstruct shredded paper for a $50,000 prize.
ONR MMOWGLI	Online multiplayer game used by the U.S. Navy Office of Naval Research (ONR) and other U.S. government agencies to perform online war games to study various problems and hypothetical scenarios.
DARPA Cyber Fast Track	Accelerated proposal process for cyber-related projects.
In-Q-Tel (IQT)	Nonprofit VC arm of the CIA to equip CIA and other agencies with the latest in information technology. About twelve investments a year.
Red Planet Capital	Nonprofit VC arm of NASA to promote private-sector participation; no longer operating.
OnPoint Technologies, also known as AVCI	Nonprofit VC arm of the U.S. Army. A handful of investments per year.

Notable in table 2 are NeedipeDIA and Challenge.gov, both platforms on which government agencies solicit ideas from a wide body of participants. These platforms, which have been online only for a few years, are generally considered successful. NeedipeDIA links back to traditional U.S. government contracting forums for participants to submit ideas. Challenge.gov works similarly to Innocentive (described above) and related platforms. The list also includes instances of one-time and recurring open innovation "challenges" to give a fuller sense of the variety of approaches the U.S. government has pursued. Let's look at these examples in more detail.

DARPA has initiated a number of prize-based competitions of varying scope and length. The DARPA Grand Challenges include 2004 and 2005 challenges, as well as the 2007 Urban Challenge, 2012 Robotics Challenge, and 2013 FANG Challenge. Each of these involved $1+ million awards. In contrast, the smaller-scale DARPA Shredder Challenge had a $50,000 prize.[21] Another DARPA challenge with a similar prize ($40,000) was the Network/Red Balloon Challenge in 2009. In this challenge, in one day teams had to locate ten red balloons placed across the United States. The winning team from MIT leveraged a unique incentive structure to encourage participants from around the country to join the team and help find the balloons.

The Massive Multiplayer Online War Game Leveraging the Internet (MMOWGLI) is an online platform established by the ONR to facilitate the war-gaming process. This platform is dramatically different from the above DARPA challenge examples. The objective is to obtain actionable scenarios for the U. S. Navy to use to change its strategies. The platform is online, and much of the data collected during the execution of the game is openly available. However, the "defense"-oriented games are generally blocked from public access.

Cyber Fast Track (CFT) was a unique DARPA program that encouraged individuals and small businesses in the cyber community to work with the U.S. government. In 2010 DARPA hired a prominent hacker to lead the program, Peiter Zatko, known by his hacker name, Mudge. The program was an experiment in leveraging creative contracting vehicles to encourage new participants and reduce contracting time. Mudge said, "DARPA isn't an open organization. We were looking for a new way to work with people."[22] DARPA awarded over one hundred contracts in under two years and achieved a "mean of 7 days for approval."[23] This was achieved through the use of an intermediary, as discussed further in chapter 4 below.

Strategy	Instance	Problem Selection	Idea Generation	Commerc-ialization	Financing	Facilitator	Solution Selection
U.S. Commercial Sector							
VC Arm	Google Ventures	Start-up	Start-up	Start-up	Google	Google Ventures	Google Ventures
	Sandisk Ventures	Start-up	Start-up	Start-up	Sandisk	Sandisk Ventures	Sandisk Ventures
	Dell Ventures	Start-up	Start-up	Start-up	Dell	Dell Ventures	Dell Ventures
U.S. Government							
VC Arm	In-Q-Tel (IQT)	Start-up	Start-up	Start-up	Intel. Comm.	IQT	IQT
	OnPoint	Start-up	Start-up	Start-up	Army	OnPoint	OnPoint
	Red Planet Capital	Start-up	Start-up	Start-up	NASA	Red Planet	Red Planet

FIG. 7. Functional decomposition of VC-arm approach in the U.S. commercial sector and the U.S. government. Created by the author.

The U.S. government appears to be following the U.S. commercial sector in the choice of design variables related to open innovation approaches. For example, the government application of the VC-arm strategy mirrors that in the commercial sector. In-Q-Tel is the VC arm of the CIA. The CIA was the first agency to use the VC-arm strategy. This strategy was mirrored by AVCI–OnPoint Technologies and Red Planet capital. U.S. government implementations of the VC-arm strategy have not significantly innovated on the model as implemented in the U.S. commercial sector. A large part of the U.S. government's justification for forming In-Q-Tel, OnPoint Technologies, and Red Planet Capital was to mimic the successes of strategic VC arms in the U.S. commercial sector.[24] This is further illustrated by the mirrored functional decomposition analysis, shown in figure 7.

When Chesbrough coined the term "open innovation" in 2003, he defined it in a private-sector context.[25] The general trend with the U.S. government has been to pursue open innovation approaches only after they have been leveraged successfully in the private sector. Crowdfunding is a good example; it is still relatively new and in formative stages in the U.S. commercial sector, and there are no substantive examples of the government's employing the approach—but it cannot be ruled out.

Imagine the U.S. government inviting the public to propose

ideas/projects to a crowdfunding platform for projects relevant to multiple government departments and agencies. The involvement of multiple government entities could keep each one's individual contribution to the overall funding of a project relatively small, with the potential for circumventing some of the contractual and bureaucratic hurdles associated with traditional acquisition processes. A government agency might be incentivized to contribute some of its budget to an initiative because it sees the innovation as having specific value for its agency mission or objectives. Further, specific incentives could be established to influence innovators' designs.

Still, as attractive as this imagined scenario might be, successful application of these innovation strategies in the U.S. government R&D context could face a challenge that does not exist in the commercial sector: *security*. The unique characteristics of *secure* U.S. government R&D environments may constrain or impede open technology innovation, thereby hampering the government's ability to enable, accelerate, and enhance innovation in areas of importance to national security.

Chapter 2 explores this security and secrecy in detail.

TWO

The World of Secret U.S. Government R&D

At the federal level the U.S. government's budget for R&D supports a broad range of scientific and engineering activities with purposes including securing the national defense, advancing knowledge generally, developing the workforce, and strengthening U.S. innovation and competitiveness in the global economy.[1] The U.S. government categorizes its R&D activities into three categories: basic research (known in the government as "6.1"), applied research ("6.2"), and development ("6.3").[2] The government also stipulates the measure of the maturity of a technology based on technology readiness level (TRL). TRL is a scale from one to nine, with one being the least mature technology and nine being a fielded system, shown in figure 8. These levels correspond to the phases of technology development from basic technology research through development to operations and support. In general a low TRL of one or two corresponds to 6.1 R&D; a TRL of three or four corresponds to 6.2 R&D, and so on. Figure 8 shows how R&D funds in the 6.1, 6.2, and 6.3 categories map to TRLs.

The U.S. government's acquisition process is unique and complex, and it involves several steps: identifying priorities and needs, soliciting proposals, establishing contracts with defense contractors, and advancing technology through the phases of technology development.[3] Because of this complexity, the government has dedicated acquisition agencies that are paired with agencies and military branches.[4] The "milestone decisions," shown in figure 8 as triangles labeled A, B, and C, are formal reviews and decision points in the acquisition process. For example, a technology must be developed and demonstrated to have achieved TRL 6 as

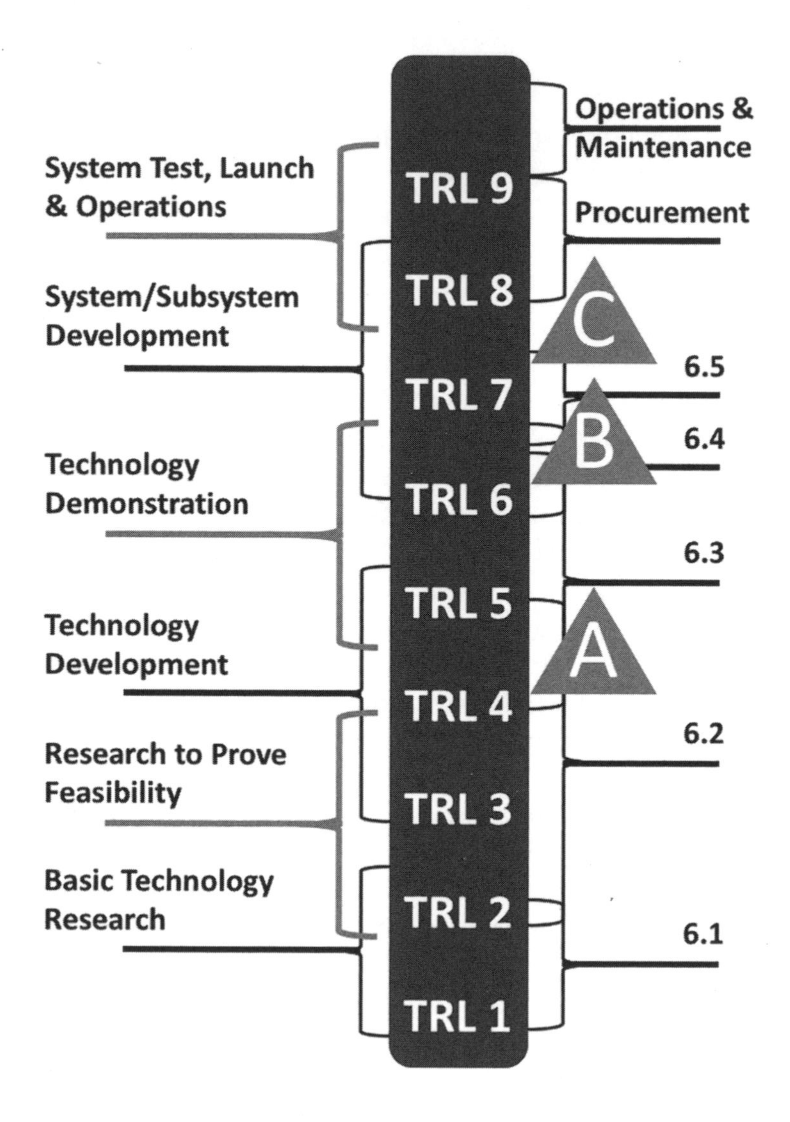

FIG. 8. Technology readiness levels and types of R&D mapped to phases of technology development. Source: Adapted from Jacobsen, "Technology Readiness Levels Introduction," and NASA and U.S. Army Medical Department Medical Research and Materiel Command, *Army Technology Objectives* (ATOS).

a prerequisite to Milestone B. A defense "program" is considered to have officially started once it passes Milestone B.

To define the scale of U.S. government R&D, it is important to understand the size of the federal budget, which is proposed by the executive branch and then must be approved by Congress. Congress defines the R&D priorities for the country and determines the "aggregate, agency, and programmatic R&D funding."[5] President Trump's 2018 proposed R&D budget was $151 billion, a little more than half of which—$84 billion—was for defense.[6] This was on a similar scale to spending levels for the last few years.[7]

The landscape of U.S. government R&D consists of many players, agencies, and organizations. Figure 9 maps the key players, including federal and state governments and nongovernmental players. The figure shows funding channels within government agencies, as well as funding for those working on U.S. government R&D contracts. Organizations outside of the U.S. government that provide funding to these contractors are also shown. In essence the figure depicts a stakeholder value network and uses arrows to show flows, covering the key categories of stakeholders relevant to U.S. government R&D environments. Each category includes a few example stakeholders.

There are three "flows" in the figure: political flows, which include approving budgets, lobbying representatives, ensuring compliance with regulations, creating or establishing courts, and providing oversight—such as that for security clearances; financial flows, which include equity investments, grants, contracts, and interdepartmental fund transfers; and the flow of technology or technology innovation.[8]

Note that many U.S. government agencies relevant here fall within federal executive branch departments headed by Cabinet members. There are also independent agencies that, while still part of the executive branch, are (or at least have traditionally been) insulated from short-term political pressure because the president's power to replace the head of such an agency is limited through congressional statutes; these include agencies such as the Central Intelligence Agency (CIA) and National Science Foundation.

Certain programs cut across the landscape in figure 9. For exam-

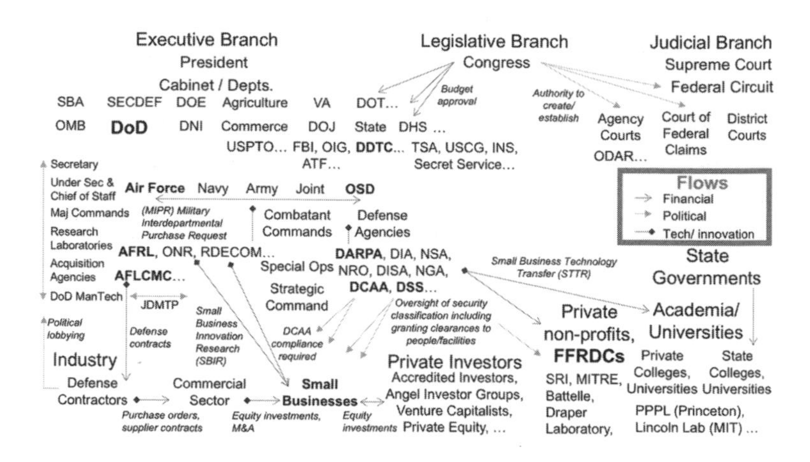

FIG. 9. Stakeholder value network: landscape of U.S. government R&D. Created by the author.

ple, Congress established the SBIR program as part of legislation that mandates eleven federal agencies to reserve a portion of their R&D funds for small businesses (defined as having fewer than five hundred employees).[9] The Small Business Administration serves as the coordinating agency for the eleven agencies, reviews SBIR program implementation, and reports annually to Congress.[10]

To illustrate how all of this works, a hypothetical electronic warfare program is discussed below, and key stakeholders are identified in bold in figure 9. This hypothetical program is led by the Air Force Materiel Command (AFMC), a major command in the U.S. Air Force (USAF) that focuses on weapon systems, receives congressional budget approval for its programs, and is overseen by the secretaries of the air force and defense, as well as—ultimately—the president. The Air Force Life Cycle Management Center (AFLCMC), an acquisition agency, oversees the identification and selection of defense contractors; this work is managed by a program executive officer (PEO).

Organizations within the Office of the Secretary of Defense (OSD) oversee various program functions, including those with a significant impact on innovation. The Defense Contract Audit

Agency (DCAA) conducts financial audits and supports contract negotiations and administration for defense agencies. The Defense Security Service (DSS) provides oversight of security classification, including the granting of clearances to individuals and facilities, while the State Department's Directorate of Defense Trade Controls (DDTC) oversees adherence to International Traffic in Arms Regulations (ITAR).

There are ways for defense agencies to acquire technology from each other simply and relatively quickly without putting new contracts in place. For instance, if the hypothetical USAF electronic warfare program seeks to leverage technology being developed at DARPA, the USAF can use a military interdepartmental purchase request (MIPR) to transfer money to DARPA. DARPA, in turn, can use the funding for programs that involve small businesses, universities, and Federally Funded Research & Development Centers (FFRDCS).

The hypothetical electronic warfare program may leverage technology developed through SBIR contracts. The USAF SBIR program is led out of the Air Force Research Laboratory (AFRL). Small businesses receive SBIR contracts from AFRL, and these are also overseen by organizations such as DCAA. Small businesses also receive funding from private investors.

Secure U.S. Government R&D

Secure U.S. government R&D is a subset of U.S. government R&D that specifically addresses national security. It tends to occur in a unique environment that restricts the disclosure of information, based on the assumption that secrecy is required to maintain national security. It probably comes as no surprise that secrecy is considered part and parcel of the United States' most advanced technology for national security. Even the CIA website's career page (shown in fig. 10) implies that "advanced technology" and "classified" are synonymous. To its credit, the CIA does have a sense of humor about its level of secrecy (see fig. 11).

In 2010 the *Washington Post* published "Top Secret America," a series of articles by Dana Priest and William M. Arkin based on a two-year investigation. Some of the key statistics reported in the

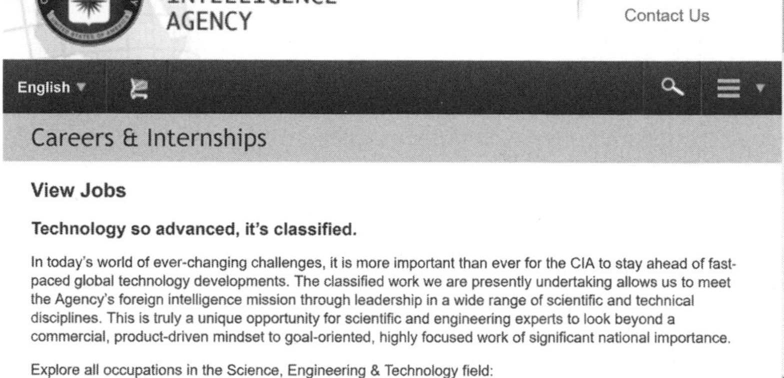

FIG. 10. CIA website careers page. Source: U.S. Central Intelligence Agency, "Careers and Internships."

articles speak to the size and characteristics of overall secure U.S. government R&D:

> The top-secret world the government created in response to the terrorist attacks of Sept. 11, 2001, has become so large, so unwieldy and so secretive that no one knows how much money it costs, how many people it employs, how many programs exist within it or exactly how many agencies do the same work. . . .
>
> Some 1,271 government organizations and 1,931 private companies work on programs related to counterterrorism, homeland security and intelligence in about 10,000 locations across the United States. . . .
>
> Every day across the United States, 854,000 civil servants, military personnel and private contractors with top-secret security clearances are scanned into offices protected by electromagnetic locks, retinal cameras and fortified walls that eavesdropping equipment cannot penetrate. . . .
>
> Every one of these buildings has at least one of these rooms, known as a SCIF, for sensitive compartmented information facility. Some are as small as a closet; others are four times the size of a football field.
>
> SCIF size has become a measure of status in Top Secret America, or at least in the Washington region of it. "In D.C., everyone talks SCIF, SCIF, SCIF," said Bruce Paquin, who moved to Florida from the Washington region several years ago to start a SCIF construction business.[11]

FIG. 11. CIA's first official tweet, June 2014. Source: CIA through @CIA Twitter handle.

It turns out that these numbers represent only a *fraction* of secure U.S. government settings. "The *Post*'s online database of government organizations and private companies was built entirely on public records. The investigation focused on top-secret work because the amount classified at the secret level is too large to accurately track."[12] The secure U.S. government R&D environment may be largely invisible to the general public, but it is vast and widespread.

Security Classification

What gets classified in the United States? Who decides what is classified? And who owns classified information? Since 1940 presidents have issued executive orders establishing U.S. government–wide classification standards and procedures.[13] As of this writing, the current order in effect is President Obama's Executive Order 13526; as in prior orders issued by other presidents, it defines three levels of classification: "Top Secret" is "applied to information, the unauthorized disclosure of which reasonably could be expected to cause exceptionally grave damage to the national security"; "Secret" applies to information whose disclosure could cause "serious damage" to national security; and "Confidential" is information whose disclosure could cause "damage" to national security.

Additional terms are used for certain classified information that gets special handling. Particularly sensitive information, designated as "Special Access Programs" (SAP) and "Sensitive Compartmented Information" (SCI), is not shared with individuals not working on that particular SAP or SCI program, even if those individuals have secret or top secret security clearances. A determination must be made that an individual has a "need to know" to gain access to information in a SAP or SCI program, and additional background checks or polygraph tests may be required. Some SAP and SCI programs are so sensitive that even top personnel in the U.S. government may be unaware of their existence; these programs are "unacknowledged," and no reference is made to them in the published federal budget.

Transmitting and handling information at the three classification levels can be cumbersome. The information must be stored in locked safes, cannot be emailed or posted online, and usually cannot even be carried by hand from one location to another without advance notice, filled-out paperwork, and the involvement of security and document control personnel at both locations. Sometimes classified information can be communicated via special fax machines and secure phones, but the lengthy process required to get approval for such technologies means that they are often outdated and cumbersome to use once they've been installed. Even more restrictions apply for SAP and SCI environments.

Only a limited number of individuals who have what is called "original classification authority" (OCA) are empowered to decide whether information should be classified and at what level: the president, vice president, agency heads, and other officials designated by the president. In specific instances other U.S. government officials are delegated OCA pursuant to very specific restrictions. OCA is used to create security classification guides (SCGs), which define what specific information is classified and at what level. Derivative classification authority (DCA) is a much broader power and is widely used—for example, by employees of defense contractors to generate classified technical documents and mark them with the appropriate classification as per the relevant SCG. Individuals exercising DCA must be trained according to procedures defined by agencies in response to executive orders. For example, a defense contractor employee working on a classified project is required to read and undergo training on the SCG; then, when generating a technical document, he or she must mark the document according to the SCG. Security personnel of both the defense contractor and the U.S. government often review these markings.

Even though SCGs include directives to classify certain information at the highest level, the SCGs may themselves be unclassified. For example, the AFRL published an SCG for electronic warfare technology in 2005 specifying that certain information was unclassified—including "acknowledgement of the use of laser jamming or sensor damage techniques" in "technical operating concepts"—but classifying at the "secret" level the "details/techniques (such as waveforms, power/fluence, etc.) used against visual or IR imaging systems."[14] The 2005 SCG specifies additional information, such as declassification instructions and the reasons for classification. The three reasons for classification cited in this SCG are the following:

- Category 1.4a: Military plans, weapons, or operations.
- Category 1.4e: Scientific, technological, or economic matters relating to the national security, which includes defense against transnational terrorism.
- Category 1.4g: Vulnerabilities or capabilities of systems, installations, projects, plans, or protective services relating to the national security, which includes defense against transnational terrorism.[15]

The categories refer to the executive order on classification, which specifies the categories of information that can be classified.

Secrecy Orders

The U.S. Patent and Trademark Office (USPTO) also plays a role in classifying information by screening patent applications to determine whether disclosure of their subject matter would have a detrimental impact on national security. This practice began during World War I, when Congress legislated that the president could withhold a patent from being issued for national security reasons during wartime.[16] The wartime restriction has since been removed. Also it is not just the president who can decide to withhold a patent; if in the opinion of an agency official the disclosure of an invention would be detrimental to national security, the commissioner of patents must enact a so-called "secrecy order." Such an order not only forbids issuance of the patent, but also forbids the inventor from publishing or disclosing any material related to the invention. Congress stipulates the actions by the USPTO in the law.[17]

There are three types of secrecy orders. The first applies to patent applications that contain unclassified subject matter but whose distribution outside the United States is restricted.[18] The second applies to patent applications that contain classified subject matter and for which appropriate controls have been identified (using DD Form 441). The third applies to patent applications that contain classified subject matter but for which appropriate controls have not yet been identified; in such cases the USPTO itself is responsible for prohibiting or approving disclosure of the invention. Sometimes when a secrecy order is imposed on a private inventor, it is referred to as a "John Doe" secrecy order.

The commissioner of patents reviews secrecy orders annually to determine whether they can be lifted; this review is waived during wartime and national emergencies. As figure 12 shows, the number of secrecy orders has increased over time. Thus the rate at which new secrecy orders are being imposed exceeds the rate at which secrecy orders are being lifted. The largest increase in secrecy orders in the last few years has come from the U.S. Navy (fig. 13).

Access to Classified Information

The U.S. government owns classified information and reserves the right to revoke access to it at any time. "A security clearance is a privilege, not a right,"[19] and the law stipulates punishments for improperly disclosing classified information or otherwise violating the obligation to protect it.[20]

For an individual to have access to classified information, he or she must at a minimum have a security clearance. In most cases security clearances are given only to U.S. citizens who are sponsored for clearance by a U.S. government agency and who undergo

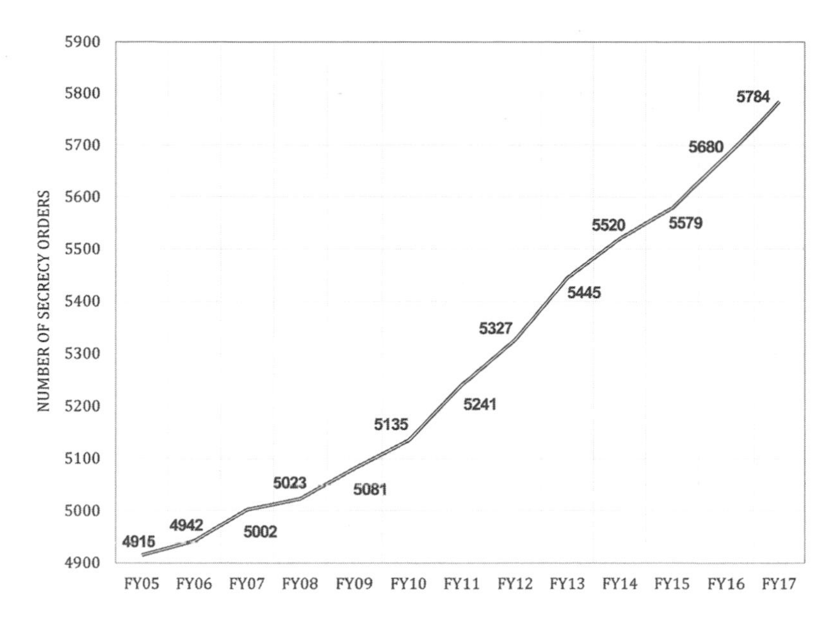

Fig. 12. Increase in total U.S. secrecy orders over time. Created by the author using data from Aftergood, "Invention Secrecy Statistics."

a security investigation to determine whether they are "of unquestioned allegiance to the United States."[21] Before individuals gain access to classified information, they are required to sign nondisclosure agreements with the U.S. government that are lifelong obligations to protect classified information—even after their security clearances are no longer active. Security clearances must be renewed every five years, a procedure that typically requires a justification statement from a U.S. government agency for continued access and another security investigation.[22]

When U.S. government agencies award a contract to a company that has not previously done business with the U.S. government, they can recommend that employees of the company be given security clearances—even if such clearances extend only to gaining access to secure government facilities and not to the handling or safeguarding of classified material at the company offices. The recommending agency becomes the sponsoring agency necessary for DSS to begin the process of investigating and granting security clearances.[23]

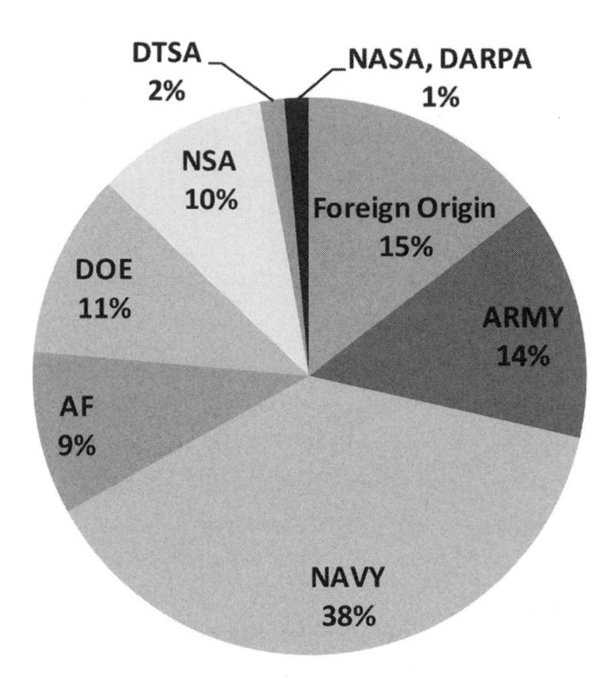

FIG. 13. New secrecy orders by U.S. government agency imposing the order. Created by the author using data from Aftergood, "Invention Secrecy Statistics."

How, one might reasonably wonder, could innovation *ever* prosper in a classified world so restricted, so siloed, and so different from the open, collaborative world we imagine, say in Silicon Valley, where high-tech innovation seems to be a daily occurrence? In fact the secrecy and restrictions do kill innovation some of the time, but there are also occasions when innovation not only survives, but also thrives.

THREE

Success and Failure in Secret U.S. Government Technology Innovation

We have seen how the desire for secrecy has driven a proliferation of restrictions and regulations that are now pervasive in secure U.S. government R&D environments. Even so, the classified world has generated tremendous innovation success. At the same time, the cloak of secrecy and its attendant web of restrictions has impeded success in some programs. To better understand what helps and what hinders secure U.S. government innovation, this chapter considers two examples: (1) one of the most famous success stories of innovation for national security in our history, and (2) a recent classified program that tried to leverage open innovation but struggled without sufficient support from the regulatory regime.

Innovation can thrive in secure U.S. government R&D environments, but it requires robust support from the regulatory regime that governs it. One of the most successful innovation challenges in world history—NASA's Apollo program—took place at the height of the Cold War in the most secretive of settings. Despite the secrecy involved, the Apollo program yielded some of the most important innovations the world has ever seen. Because of Apollo, the United States not only advanced human space exploration, but also achieved broader technology advancement in many other fields, including more than two thousand spinoff products as of 2019, from kidney dialysis machines and home insulation to flame-resistant textiles and modern athletic shoes.[1]

In the fifty years since Apollo the essential mission of secure U.S. government R&D has not changed. Over that same period innovation strategy itself has evolved and advanced along the lines discussed in chapter 1. The evolution and advancement of regulations surrounding classified environments, however, have not kept pace—and innovation in classified environments has been hindered. While the commercial sector leverages new innovation strategies to great success, the restrictions endemic to secure U.S. government R&D persist, putting U.S. national security at risk.

A Man on the Moon

One of the most famous, and perhaps the greatest, technological feats of our history was landing a man on the Moon. The Apollo 11 mission is a national achievement that marked our success in cultivating unprecedented technology innovation from R&D investment. It is also an illustration of the power of using technological superiority to achieve national security goals.

The Apollo program originated in the midst of the Cold War. In their arms race, the United States and Soviet Union were increasing their ballistic missile arsenals. Bomb shelters and nuclear air raid drills were common throughout the United States in the 1950s and early 1960s.[2] In schools so-called spotters "were assigned to watch the skies for anything that looked suspicious or out of the ordinary."[3]

During this period both the United States and Soviet Union were developing space satellite technology. In 1957 the Soviets successfully launched the Sputnik satellite, catching the United States by surprise. The first artificial Earth satellite, Sputnik was visible from all around the globe. "Set against the backdrop of the Cold War tension between the Soviet and Western blocs, both sides' preoccupation with nuclear weapons technology, and the recent conflict on the Korean peninsula, it is not difficult to imagine the fear and consternation that this technological trump card engendered in Western countries."[4]

The Soviet launch precipitated the Sputnik crisis—a period of uncertainty, fear, and so-called "Sputnik shock" in the United States. The combination of fear and surprise was a powerful force that drove U.S. national security priorities and the nation's space strategy. Sputnik's launch triggered the "Space Race" between the two countries. President John F. Kennedy supported the creation of DARPA (at the time, ARPA), whose founding mission was to "prevent and create strategic surprise."[5]

The United States' global leadership position had been threatened—and the country responded. The United States had failed to be first with a satellite in space, but the country was determined not to come in second again. Yet in April 1961, just four years after the Sputnik launch, Soviet cosmonaut Yuri Gagarin became the first person to fly in space. This achievement only heightened U.S. fears. Less than a month later President Kennedy issued a historic challenge to the nation to land a man on the Moon and return him safely to Earth. President Kennedy elaborated on this challenge in a famous speech at Rice University in Houston, Texas, on September 12, 1962:

> The exploration of space will go ahead, whether we join in it or not, and it is one of the great adventures of all time, and no nation which expects to be the leader of other nations can expect to stay behind in the race for space....
>
> Those who came before us made certain that this country rode the first waves of the industrial revolutions, the first waves of modern invention, and the first wave of nuclear power, and this generation does not intend to flounder in the backwash of the coming age of space. We mean to be a part of it—we mean to lead it. For the eyes of the world now look into space, to the Moon and to the planets beyond, and we have vowed that we shall not see it governed by a hostile flag of conquest, but by a banner of freedom and peace. We have vowed that we shall not see space filled with weapons of mass destruction, but with instruments of knowledge and understanding.
>
> Yet the vows of this Nation can only be fulfilled if we in this Nation are first, and, therefore, we intend to be first. In short, our leadership in science and in industry, our hopes for peace and security, our obligations to ourselves as well as others, all require us to make this effort, to solve these mysteries, to solve them for the good of all men, and to become the world's leading space-faring nation.[6]

Thus President Kennedy made clear that demonstrating technological superiority in space was "instrumental for U.S. security reasons."[7] To that end NASA's R&D investment increased from $89 million in 1958 to $5.9 billion in 1966—an all-time high as a percentage of the federal budget—to enable the technology innovation needed to support technological superiority in space.[8]

There was a sense of urgency to leverage strategies that could accelerate innovation. President Kennedy promised to safely land a man on the Moon before the decade was over. The United States succeeded. On July 20, 1969, Neil Armstrong and Buzz Aldrin stepped onto the Moon and planted an American flag.

Seven years may seem like a long time to achieve such a feat, but the technological requirements and system complexity of landing a human safely on the Moon and transporting him or her safely back to Earth are still, to this day, among the greatest challenges humankind has ever taken on. As President Kennedy put it that day in Houston:

> But if I were to say, my fellow citizens, that we shall send to the Moon, 240,000 miles away from the control station in Houston, a giant rocket more than 300 feet tall, the length of this football field, made of new metal alloys, some of which have not yet been invented, capable of standing heat and stresses several times more than have ever been experienced, fitted together with a precision better than the finest watch, carrying all the equipment needed for propulsion, guidance, control, communications, food and survival, on an untried mission, to an unknown celestial body, and then return it safely to Earth, reentering the atmosphere at speeds of over 25,000 miles per hour, causing heat about half that of the temperature of the sun—almost as hot as it is here today—and do all this, and do it right, and do it first before this decade is out—then we must be bold.[9]

MIT's Instrumentation Laboratory (which later became the Charles Stark Draper Laboratory) won the first Apollo contract: to develop the guidance navigation system for the Apollo spacecraft.

The cutting-edge technologies developed to enable the Apollo mission were disruptive. In the space technology area, they paved the way for satellite communications, GPS navigation, surveillance, and intelligence that are still necessary for critical military applications.[10] Technologies developed in support of the Apollo mission also had a significant impact in areas beyond space technology: "technologies like breathing apparatuses, fabric structures, communications and protective coatings that made man's step on the Moon possible soon led to giant leaps in technology on Earth."[11] Advances in rocketry, civil engineering, electrical engineering, and aeronautical engineering that support the military but are also used in civilian applications can all be credited to the Apollo program. Thus not only did the Apollo mission achieve the national security goal set out by President Kennedy, it also propelled industry forward in the country, boosting national competitiveness in many other areas.

To be clear, it would be wrong to characterize the success of the Apollo program as unqualified. It was canceled before Apollo 18 flew (the final Apollo missions, 19 and 20, were already planned at that time), and the heavy-lift technology that had been developed for the Saturn rockets that launched the Apollo missions remained unused after the program. Still the successful Moon landing is not only a source of national pride, it was also vital to U.S. national security interests during the Cold War. The achievement also showcased what is possible when appropriate support is behind secure U.S. government R&D.

As we will see in the next example, however, the lack of that support from the regulatory structure may impede success. Innovation thrived in Apollo's classified security environment in part because of robust support from the regulatory regime. Over the last fifty years, however, the regulatory regime governing classified security environments has not kept pace with innovation strategy, creating missed opportunities. The story of FANG is a story of just such a missed opportunity, in which decades of restrictions and regulations could not adequately accommodate the open innovation strategies a classified program hoped to leverage to achieve technological advancement for national security.

The U.S. Government Tries Classified Open Innovation

In the 1990s the U.S. Marine Corps identified a growing national security threat to amphibious and coastal operations. Although U.S. amphibious operations were critical to victory in both the European and Pacific theaters of World War II, the global proliferation of mines, missiles, and artillery since then "force [the

Marine Corps] to travel farther . . . and faster if the Marines are to avoid threats and successfully reach the beach."[12] The Marines require an infantry vehicle that is amphibious; it needs to be fast and maneuverable on both land and water. But the Corps' aging legacy vehicle—called the Amphibious Assault Vehicle (AAV)—is more than four decades old, having gone into operation in 1972. Problems have persisted: "With every year that passes, the waterborne armored vehicle becomes more vulnerable, less reliable, and potentially more accident-prone."[13] According to the U.S. Government Accountability Office, "As weapons technology and the nature of threats have evolved over the past four decades, the AAV is viewed as having limitations in water speed, land mobility, lethality, protection, and network capability."[14]

Because the AAV was in need of a replacement, the government tried to develop a vehicle that would meet the Marines' requirements. Why is this so technologically challenging? When such a vehicle is on land, it must be "rugged and survivable." Yet armor that is rugged enough to survive mines, missiles, and artillery has so far been "too heavy for sea-based assaults" and interferes with the requirement to "transport troops . . . at high speeds."[15] These seemingly conflicting requirements clearly pose a bold challenge!

No matter what it tried, the government could not crack the nut. Unable to solve the problem using traditional innovation methods, the government turned to DARPA in 2012 to find a new way to solve the problem. DARPA is a unique entity within the U.S. government that is known for pushing technology limits and using creative contracting to support innovation. DARPA's mission, as discussed above, is to create and prevent strategic surprise, and it was established in response to the strategic surprise of the Soviet launching of Sputnik, which threatened U.S. national security. DARPA has been called "the Pentagon's 'think outside the box' research arm" and sees itself as the "DOD's primary engine for technology innovation."[16]

To solve the U.S. Marines' challenge, DARPA decided to try open innovation because of its demonstrated success in achieving technological breakthroughs in the commercial sector. To do so DARPA created a program called the Fast Adaptable Next-

Generation Ground Vehicle (FANG). The FANG program set out to use the Challenge/Competition open innovation strategy. While this was not DARPA's first foray into open innovation, FANG was unique because the technical requirements for FANG are classified. DARPA FANG leadership has noted that FANG is the first and only U.S. government program to date that sought to tackle secrecy issues in the context of open innovation—a point of considerable pride.

Under the leadership of DARPA director Regina Dugan and DARPA program managers Lt. Col. Nathan Wiedenman and Paul Eremenko, FANG was originally devised as a three-tier project beginning with "FANG-1" in order to address the secrecy issues. For example, the specific requirements on how rugged armor must be in order to survive mines, missiles, and artillery are classified. Therefore the development of the chassis and armor design was reserved for FANG-2, which contemplated the potential for secret-level submissions. The drivetrain was identified as an important vehicle component with the least number of classified requirements, and therefore it was chosen for the FANG-1 challenge. Yet even FANG-1 faced challenges due to secrecy.

FANG-1 was a three-month challenge run in 2013 focused on designing a drivetrain for an amphibious vehicle to support the U.S. Marines. About one thousand innovators participated, organized into about two hundred teams. While a winning team was selected and prize money was awarded, many of the design requirements remained unsolved. Ultimately FANG-2 and FANG-3 were canceled. Before we turn to some of the fundamental secrecy challenges that hindered the FANG program, including tensions within the regulatory structure, it is worth discussing the non-secrecy-driven open innovation design choices. While the program was ultimately canceled, FANG did achieve important advances in the state of technology and opened the door for further study of the potential for open innovation in secure R&D environments.

FANG-1 Open Innovation Design Choices

Figure 14 shows the FANG-1 design variable choices for the Challenge/Contest/Game strategy (discussed in chapter 1). Now that we

Design Variable	Options	VC Arm	Challenge/ Gamification	Innovator Network	Crowd Funding
Objective is to get more money invested in solving the problem than just what you have to spend	Yes No	Yes No	Yes No	Yes No	Yes No
Objective is to increase general/public awareness of the problem/need	Yes No	Yes No	Yes No	Yes No	Yes No
Objective is to lower the threshold for participation/contributing to solving problem	Yes No	Yes No	Yes No	Yes No	Yes No
Objective is to get user/customer buy-in early	Yes No	Yes No	Yes No	Yes No	Yes No
Recurrence Model	O – One-time T – Tiered P – Platform	O T P	O T P	O T P	O T P
IP protection offered	N – No M – Medium Y - Yes	N M Y	N M Y	N M Y	N M Y
Commitment/Investment required to participate (financial, duration of competition, # participants, Deliverable).	S – Small M – Medium L - Large	S M L	S M L	S M L	S M L
# of winners	Z – Zero F – One/Few M – Many N – N/A	Z F M N	Z F M N	Z F M N	Z F M N
# participants or # of competitors	H – Hundreds T – Thousands M - Millions	H T M	H T M	H T M	H T M
Individual/team. Team size	I – Individual S – Small C - Consortium	I S C	I S C	I S C	I S C
Prize Offered	M – Monetary Prize N – Non-Monetary Prize X – No Prize	M N X	M N X	M N X	M N X
Prize distribution	F – First Place Only T – Top Few E – Everyone N – N/A	N	F T E	F T E	F T E
Participant (school, individual, community, supplier, user)	S – School I – Individual C - Company	S I C	S I C	S I C	S I C

FIG. 14. FANG-1 open technology innovation approach. Created by the author.

have the benefit of hindsight, it is worth considering how changing the design choices for FANG-1 could have affected the open technology innovation approach.

Let's consider the possible effects of changing the size of the monetary prize. FANG-1 awarded a $1 million prize, and to do that DARPA needed approval from Undersecretary of Defense Ashton Carter, in charge of acquisition, technology, and logistics (AT&L), to delegate the prize authority of the director of defense research and development directly to the DARPA director. Obtaining prize authority secured for DARPA more autonomy over the choice of design variables with the FANG-1 challenge and enabled DARPA to make prize awards directly to the winners. High prize amounts like FANG-I's tend to attract a highly trained subset of the population and incentivize participants to put in the required effort.

Let's compare FANG-I with a predecessor program, the 2011 eXperimental Crowd-Derived Combat-Support Vehicle (XC2V) design challenge. XC2V had as its purpose to "learn more about the crowdsourcing process."[17] Unlike FANG, the XC2V challenge did not involve sensitive information or classified requirements. It did, however, help point the way forward for open innovation and involved a community of participants that grew to nearly seven thousand. XC2V gave one hundredth the prize amount as FANG-I but attracted a larger section of the population. One potential downside of having a single large prize is that it may discourage collaboration as teams compete to win. Indeed FANG-I participants tended to be competitive and kept their designs to themselves, whereas XC2V participants tended to be more collaborative and openly shared their designs with other members of the community, seeking feedback.

FANG-I and other instances of Challenges in the commercial sector have required a high commitment/investment by participants, a variable that could have been moderated to achieve a different outcome. Decreasing required commitment could have brought in additional innovators with more varied expertise, but it may also have necessitated a less specialized design challenge. Increasing the required commitment could have involved expecting some prototyping and/or lengthening of the challenge duration, but it might also have required increased incentives or targeting a different audience.

FANG could also have simplified its objectives. FANG sought to design an amphibious vehicle for the Marines, but there was an additional objective: test DARPA's META software development toolchain. Therefore, FANG required innovators to use certain tools during the innovation process. FANG did achieve the objective of testing META, but this additional, competing objective detracted from innovating on the drivetrain. A survey of FANG-I participants revealed that many were constrained in their innovation because the META software was reportedly immature and brittle. A FANG-I case study reported the survey results as showing "both excitement and enthusiasm, as well as frustration and disappointment regarding some aspects of the competition experience."[18] Arguably the objectives of building an innovative drivetrain and testing the META software could have been decoupled. FANG had a tiered recurrence model and offered a monetary prize, but that may not be the most conducive model for testing software. Rather than a tired recurrence model, perhaps the META toolchain could have been opened up as a platform. Further, a non-monetary prize might have cast the testing of META as a game in which the gaming experience itself could have been the prize—as is the case with Foldit. Like META, Foldit—which bills itself as a place to "Solve Puzzles for Science"—sought external participants to engage with its platform, in this case to explore how proteins could be folded. Rather than offering monetary prizes, Foldit encourages participation by publicizing in-game achievements to other players and thus creates reputational incentives.[19] Participants are further motivated through "social praise (chats and forums), . . . engaging gameplay . . . and the connection between the game and scientific outcomes."[20] Offering such non-monetary incentives may have been a way to decouple the competing FANG objectives by leveraging alternative open innovation strategy options to improve FANG's success.

How might FANG-I have performed had DARPA followed the typical VC-arm approaches rather than Challenge approaches? First, FANG-I could have accommodated a smaller number of larger, more focused teams by changing its target participant type and number. Simultaneously, by increasing the duration of the effort, the financial incentive could have been tiered even within the first phase, a step that would have allowed for incremental objectives and potential error correction by DARPA staff during the program (e.g., to compensate for immature tools). Incentives could have been offered such as a potential follow-on contract with the end customer, the U.S. Marines. A larger, more dedicated team would not have been targeting the prize money alone, but also a sustainable business model, and hence might have been more willing to work through security classification hurdles, such as by following the process of obtaining security clearances to improve the participant's understanding of the design requirements.

Challenges of Secrecy and Participation

While the above discussion highlights some of the various design variable choices with respect to open innovation strategy implementation made by FANG, it was security classification and data sensitivity issues that ultimately led to FANG's cancellation.

As discussed above, classified survivability requirements were deferred to FANG-2 by focusing FANG-1 on the drivetrain. However, even the drivetrain was subject to sensitive requirements, such as for maneuverability survivability in rough seas. While DARPA was experimenting with creative ways to open innovation for classified security needs, it had to contend with the statutory mandates imposed on the State Department, which regulates such defense technology.[21] Specifically the Directorate of Defense Trade Controls (DDTC) in the State Department implements the ITAR and controls the export of defense articles that provide a critical military or intelligence advantage. DARPA debated with DDTC about whether the FANG-1 drivetrain models were subject to ITAR, which would restrict participation and distribution to certain people and companies. Whereas DARPA was seeking to leverage open innovation to advance technology development in an area of importance to national security, the State Department was bound by legislative mandate. This is an example of tension within the regulatory regime that impeded open innovation.

A 2011 DARPA document noted that while ITAR are often "clear cut, dual-use technology can sometimes be challenging to categorize." (Dual-use technologies are those with both military and civilian applications.) DARPA promised further that it would "carefully address ITAR issues on a case-by-case basis."[22] In 2012 DARPA acknowledged that it was "likely that at least some IFV [infantry fighting vehicle] designs will fall within the scope of 22 CFR § 121, The United States Munitions List"—meaning that ITAR would apply and limit what could be done openly.[23]

It was challenging for DARPA to implement ITAR restrictions in the midst of the open innovation effort. For example, ITAR's export restrictions made it necessary for DARPA to check IP addresses to ensure participants logged in only from the United States. Limiting innovators to those from the United States reduced the pool of participants. It should be noted that more recent 2017 legislation restricts all government prize competitions to U.S. persons or companies.[24]

The pool of participants for FANG-1 was likely further reduced because of its demanding IP policy. The IP policy was difficult to understand and placed the burden on potential participants to assert their rights. Separate 2017 legislation on crowdsourcing and citizen science requires the disclosure of IP policy "and other terms of use to the participant in a clear and reasonable manner."[25] Such a provision is not included in the 2017 legislation on prize competitions.

The policy in FANG-1 was also potentially perceived as less fair than some other commercial approaches because participants forfeited some IP rights to ideas even if they were not selected or rewarded with prize money (discussed further in chapter 6).[26]

As DARPA continued to contend with secrecy challenges, the specific technical requirements that the vehicle drivetrain proposals were being judged against were obscured from innovators. A survey of participants revealed that the opacity of the requirements was a hurdle to innovation. Even the winning design fell short of several objectives, including driving performance and maneuverability, and it scored only 4,613 out of a theoretical maximum of 7,307 points (63 percent).[27]

To continue FANG into the second and third tiers would have required more stringent restrictions on participation since innovators needed to address survivability requirements that were not just ITAR-controlled, but also classified. FANG-2, with a prize of $1 million, was to focus on developing the vehicle chassis and meeting classified survivability requirements. FANG-3 was to complete the total vehicle platform, with a prize of $2 million.

Even with all the challenges with FANG-1, however, the program managers were committed to the open-innovation path and wanted to "accept SECRET-level submissions" for FANG-2 and FANG-3, as figure 15 shows under "Participant Pool."[28] "Pooling the brains of 300

APPROVED FOR PUBLIC RELEASE. DISTRIBUTION UNLIMITED.

Fast, Adaptable Next-Generation GCV (FANG)

Mobility/Drivetrain Challenge

SCOPE
- Vehicle drivetrain to meet GCV speed, efficiency, terrain, reliability objective
- Available model library to include:
  - Hybrid-electric systems
  - Novel ground interfaces

PARTICIPANT POOL
- Global

INCENTIVE SCHEME
- Prize ~$0.5-1M for winning design
- Winner(s) judged based on multi-objective weighting function

DESIGN AGGREGATION
- Use of META metalanguage required
- Use of vehicleforge.mil optional

BUILD APPROACH
- iFAB foundry build for top design(s)

Chassis/Integrated Survivability Challenge

SCOPE
- Chassis and armor design to meet principal GCV-like survivability objectives
- Available model library to include:
  - Advanced armor concepts
  - Novel configs (monocoque, v-hulls)

PARTICIPANT POOL
- Global
- May accept SECRET-level submissions (?)

INCENTIVE SCHEME
- Prize ~$0.5-1M for winning design
- Winner(s) judged based on multi-objective weighting function

DESIGN AGGREGATION
- Use of META metalanguage required
- Use of vehicleforge.mil optional

BUILD APPROACH
- iFAB foundry build for top design(s)

Total Platform Challenge

SCOPE
- Complete GCV based on core Army objectives and distilled requirements

PARTICIPANT POOL
- Global
- May accept SECRET-level submissions (?)

INCENTIVE SCHEME
- Prize ~$1-2M
- Winner judged based on satisfaction of constraints and multi-attribute preference function (i.e., entirely objective approach)
- Additional/portion of prize after T&E

DESIGN AGGREGATION
- Use of META metalanguage required
- Use of vehicleforge.mil optional

BUILD APPROACH
- iFAB foundry build for top design(s)

FIG. 15. DARPA FANG tiered challenges. Source: Eremenko and Wiedenman, "Adaptive Vehicle Make (AVM)" (emphasis added).

million Americans could turn out to be a masterstroke if DARPA can traverse issues such as classified requirements."[29]

DARPA's prize authority delegation approval was narrowly limited to the "Challenge" approach since that was what the organizers had originally envisioned. During the course of FANG-1 the organizers faced significant hurdles due to the "secrecy challenge" (discussed further in chapter 5), and they considered alternative innovation strategies. However, the prize authority delegation approval was not flexible enough to enable FANG to evolve its open innovation strategy.[30]

Unfortunately FANG-2 and FANG-3 were ultimately canceled because, as DARPA indicated, "the consideration of sensitive materials and analysis [is] not suited for the public domain."[31] The program did not fulfill the promise of open innovation in the way that was hoped. The public documents regarding the restructuring of the Marines' previous program, the Expeditionary Fighting Vehicle (EFV) program, and the follow-on Amphibious Combat Vehicle (ACV) program indicate that the Marines did not ultimately adopt the FANG-1 design.[32] The U.S. Marines' need for the amphibious vehicle remained unmet. FANG's cancellation left the Marine Corps with the legacy AAV and what one writer referred to as its "survivability upgrade program"—with the Corps in 2019 "looking for new tracks [to enhance its mobility and] to prolong the life of the vehicle into 2035."[33]

The secrecy challenge FANG faced is important to the focus of this book, and the FANG story suggests opportunities to effect sustainable changes that would support the success of future efforts. Apollo and FANG offer some compelling lessons. Both started with an objective that was unachievable with existing technology. In both cases fulfilling the objective was vital to national security. Apollo shows that innovation can thrive in secure U.S. government R&D environments, despite their restrictions. Future innovation in secure U.S. government R&D needs to replicate the success of Apollo. But without the requisite support from the regulatory regime, FANG was unable to achieve its ultimate vision of enabling classified programs to leverage open innovation widely.

Apollo benefited from the support of an entire country. The space program was a national effort. It was an effort ordinary citizens embraced. Brilliant engineers and innovators in the country wanted to work on nothing else. That is a special kind of incentive that never existed with FANG, and it is not likely to exist for many, if any, U.S. government R&D programs going forward. That is precisely why openness is needed. Innovation and the participation of innovators is the way to bridge the gap between success and failure absent an Apollo-like motivation. Without the support for openness by the regulatory regime, an increasing number of our critical national security needs will go unmet.

FOUR

Practical Consequences and Perverse Incentives

Regulatory regimes—including executive orders, legal codes, agency regulations, and judicial checks and balances—come together to influence and govern multiple aspects of the innovation pipeline for national security. Even regulatory regimes that reflect a commitment to innovation often simultaneously serve other values and policies that are more or less in tension with innovation. As a result, the regulatory regimes themselves may incorporate constraints that limit innovation.[1]

In this chapter I first identify some of the values of the regulatory regime governing secure U.S. government R&D, including national security, job creation, and economic growth. Then I explore a few examples of emergent behavior and unintended consequences that result from the regime's service to multiple competing values. In the two chapters that follow I take a deeper dive into two main areas where the regulatory regime is unintentionally constraining innovation in secure U.S. government R&D settings: secrecy and participation.

System-level analysis of regulatory regimes governing secure U.S. government R&D reveals that policy tensions do exist and, indeed, that some U.S. laws and policies meant to further national security have unintended consequences. Such emergent behavior may actually be detrimental to national security. From a systems engineering perspective, the term "emergent behavior" refers to a property that appears when a number of entities operate in a system and generate complex behaviors as a collective. Emergent behavior may be unforeseen by a given entity, and an entity may even take actions that it does not realize are directly detrimental to its own primary objective.

To illustrate the concept, consider the ITAR regulatory regime with respect to the telecommunications satellite industry. In the industry's formative years the United States had a decisive lead over other nations in the development and deployment of telecommunications satellite technology. Because this technology was deemed critical to national security, the federal government shifted some regulatory authority from the Commerce Department to the State Department, which then restricted dissemination of telecommunications satellite technology to other countries using ITAR regulations in an effort to protect U.S. technological superiority. Unfortunately this protectionist move created an unintended result. It actually compelled other countries, which might otherwise have been content to rely on U.S. technology, to develop their own telecommunications satellite capabilities.[2] These other countries, facing significant roadblocks in acquiring what they felt was a necessary technology, shifted their attention inward and began developing the capabilities on their own.

The result is that the United States is not the only nation with telecommunications satellite technology. Instead international market share of telecommunications satellite technology has grown tremendously, as have international competencies in such technology. At the same time, as figure 16 shows, the U.S. share of the global market has declined dramatically in part because other countries were unwilling to navigate the complex and intentionally difficult acquisition process via ITAR to obtain U.S. telecommunications satellite technology. The United States has not recovered its market share to what it was before the 1998 ITAR regulation.[3]

To conduct a system-level examination of the regulatory regimes governing secure U.S. government R&D first requires identifying the values that may be in tension with innovation. This requires an understanding of the factors introduced in chapter 1 that support robust and open innovation: full and open competition, an environment in which innovators have access to collective knowledge and the ability to share ideas to collaborate, and a structure in which potential innovators are properly incentivized to participate.

The U.S. government has also recognized these factors as sup-

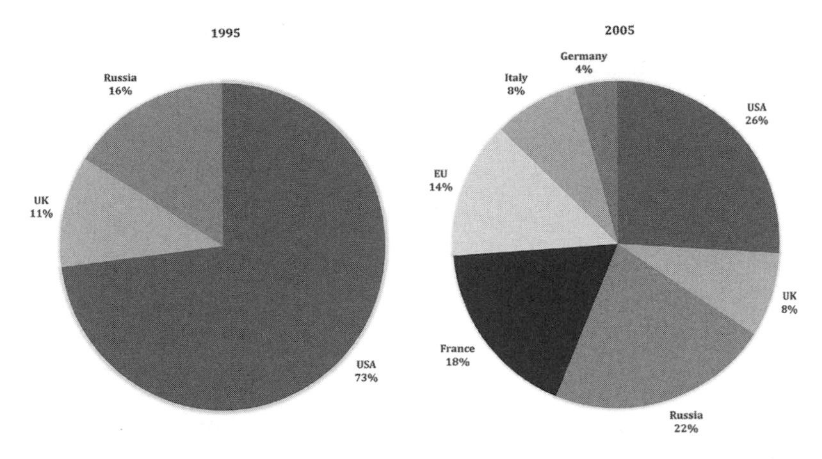

FIG. 16. Declining U.S. share of satellite exports. Source: Meijer, *Trading with the Enemy*.

porting robust and open innovation: "Competition," as the U.S. Government Accountability Office (GAO) recognizes, "is a cornerstone of the acquisition system and a critical tool for achieving the best possible return on investment for taxpayers."[4] Full and open competition "is intended to create a competitive market and enable the government to select the most innovative products and services at the lowest prices."[5]

Competing Values Served by Regulation

As noted, some values of the regulatory regimes governing secure U.S. government R&D may be, to some degree, in tension with innovation. These include not only national security, but also job creation, economic growth, fairness, rule of law, and other constitutional principles. This is not an exhaustive list, but rather a sample of values that may be most relevant here. None of these values is fundamentally at odds with innovation. These values and innovation are not mutually exclusive. Still tensions arise under the regulatory regime while competing values and priorities are being served. Let's look at these values more closely.

National security, the "national defense or foreign relations of the United States," is the primary value that is unique to *secure* U.S. government R&D.[6] In fact it is the dominant value in that environment, and, almost by definition, secure U.S. government R&D serves national security.

Agencies that conduct and oversee secure R&D have this national security objective enshrined in their missions, establishment documentation, and strategy directives. For example, the National Science Foundation Act of 1950, which established the NSF, stated its mission: "To promote the progress of science; to advance the national health, prosperity, and welfare; to secure the national defense; and for other purposes."[7] DARPA, too, incorporates national security into its mission and strategic direction:

> America's national security challenges are in many respects more complex, dynamic and widespread than ever before. For example, no longer does the United States have the luxury of focusing its defenses almost exclusively on a single superpower; today the nation faces multiple real and potential adversaries, including non-nation-states resistant to conventional diplomatic and military pressures. At the same time, sophisticated technologies once accessible exclusively to a few global powers are increasingly available and affordable. And new and serious threats . . . are emerging and evolving at a rapid pace. These trends demand that DARPA diligently pursue its longstanding mission of creating and preventing strategic surprise. [8]

A multitude of laws and policies that form part of the overall regulatory regime also directly point to national security as a primary objective. For example, President Obama's 2009 Executive Order 13526 on "Classified National Security Information" introduces the policies and procedures for classifying information with the purpose of supporting national security.[9] In the National Security Act of 1947 Congress explicitly states its purpose for the legislation as "provid[ing] a comprehensive program for the future security of the United States."[10]

While the foundational statements about agencies' objectives to serve national security seem straightforward, laws and policies designed to support national security objectives that comprise the regulatory regime sometimes create tensions with innovation. For example, directed R&D to support innovation related to a specific technical problem is hampered when the details of the technical problem must be obscured for national security reasons. I call this the *secrecy challenge*. Further, innovation strategies that depend on open collaboration to bring forth the fullest breadth of potential solutions are impeded when the regulatory regime takes these innovations in the name of national security and does not adequately incentivize key potential innovators to participate in the innovation process. I call this the *participation challenge*. In subsequent chapters both challenges will be examined in more detail and potential approaches to address them will be offered.

Job creation is another value of the regulatory regimes governing secure U.S. government R&D environments. For example, NSF R&D solicitations require submissions to address "the anticipated impact on job creation as a result of this innovation."[11] A DOD memo from 2014 lists "strengthen[ing] economic growth and job creation" as goals of its broad R&D initiatives.[12]

However, job creation and innovation are different, sometimes unrelated objectives with different metrics of success. Prioritizing job creation in the selection of R&D proposals may occasionally stifle innovation. For example, research has shown that there may be innovation advantages that come from the agility of smaller teams, yet a small company with an innovative proposal could be passed over due to unrelated challenges in growing its team.[13]

The next value, *economic growth*, refers to the growth of the U.S. economy. NSF R&D solicitations require submissions to address economic growth: "Proposals must describe a compelling business opportunity to be enabled by the proposed innovation." And proposals are required to state the "estimated size of the market being addressed."[14]

It is possible, and in some cases likely, that certain national security technology development needs do not correspond to a large market opportunity for economic growth, thus creating tension with innovation. For example, the national security need might be for a niche technology that would greatly benefit the military but have little commercial application. At least with respect to SBIR awards in U.S. government R&D, however, this tension does not appear to be strong. SBIR awards are generally granted if an innovation addresses a need regardless of the economic benefit. "Many program officers focus on making awards to firms that are most likely to meet agency needs, with little regard to the effect of awards on the pace of technological innovation or even to the long run sustainability of the awardees as businesses."[15] Still government R&D officials may feel pressure to demonstrate performance that corresponds to economic growth. "The contravening pressure exists to back firms that are likely to be successful. . . . The Small Business Administration and program officers at many agencies compile stories of successful awardees, which are frequently presented in speeches and reports. This may lead to pressures to select firms that would succeed even without the award."[16]

Finally, there are the interests of *fairness*, *rule of law*, and *other constitutional principles*, which apply across regulatory regimes—not just those that govern secure U.S. government R&D environments. These include due process, non-discrimination, equal opportunity, and others. Agencies tasked by Congress with developing and administering regulatory regimes for programs must promulgate regulations and rules to ensure these interests are protected.

In the context of the government acquisition process, the two most frequently applied bodies of regulation are the Federal Acquisition Regulations (FAR) and the Defense Federal Acquisition Regulation Supplement (DFARS). To ensure constitutional non-discrimination and equal opportunity, FAR and DFARS clauses have been developed that over time have become so long and complicated that potential participants may require a lawyer to interpret them. This need for expert personnel may have an unintended effect: disadvantaging the low-income, underprivileged groups the clauses were intended to protect.

"FAR itself," writes one researcher, "is an inhibitor to perfect competition because it is seen as a barrier to entry for small companies with limited resources or companies with no previous experience contracting with the government. The FAR contract has become so increasingly complex that companies must hire personnel with expert knowledge of the detailed requirements outlined in the FAR."[17] In other words, these clauses may actually be barriers to participation that in turn impede innovation.

Maneuvering through a Regulatory Bureaucracy

The complexity of the regulatory bureaucracy itself may lead to unintended consequences, such as deterring all but the most sophisticated players from participating. For innovators, especially small businesses or those not experienced with government contracting, maneuvering through the regulations and navigating the bureaucracy can be extremely difficult. In some cases the pace, difficulty, and cost of engagement with the government can prohibit participation altogether.

Contracting with the government is unique. It requires special qualifications and overhead for a would-be government contractor. A business must go through a comprehensive registration process in the System for Award Management (SAM) just to be considered for a government contract. U.S. government contractors are required to maintain updated information in both SAM and agency-specific systems, which are often not coordinated with SAM—adding an extra burden. All in all, a small business trying to apply for an SBIR award has to meet the burden of registration in at least four separate places: SAM, the Dun and Bradstreet Data Universal Numbering System (DUNS), the Small Business Administration's company registry, and the specific U.S. government agency to which it is applying. In the case of the U.S. government agency, both the company and the principal investigator have to register, and some agencies require a separate, additional registration.

The requirements are even more onerous for those hoping to become "defense" contractors. Applying companies must also comply with requirements from the DCAA and Defense Contract Management Agency (DCMA), a task that can be quite challenging. The government recognizes this problem and hosts special webinars (such as online seminars and video tutorials) to educate companies about what each agency does and how agencies differ. Companies are required to work with agencies before a defense contract is awarded and also during the contract performance period.[18]

Once selected for a potential award but before payment or confirmation of the award, a contractor must enter the contract negotiation process, which involves negotiating and agreeing to hundreds (if not more) of FAR and DFARS clauses that will apply to the contract. Some of the most important of these pertain to IP rights, although it can be a challenge to identify which FAR and DFARS clauses are related to such rights since they are not typically identified specifically during negotiations. The clauses are presented in one long list with no categories; clauses related to, say, employee drug testing and accounting are interspersed with those about IP. Many small businesses and start-ups cannot afford to engage legal counsel for this process. Even for large businesses and established defense contractors that do consult attorneys, the process is far from simple or intuitive.

Let's first take a closer look at the navigational challenges the defense R&D bureaucracy poses for the large defense contractors, known as the "Big Five"—the top five U.S. defense contractors in terms of size of government contracts: Boeing, Lockheed Martin, Northrop Grumman, Raytheon, and General Dynamics.[19]

To negotiate FAR and DFARS clauses in contracts, these large defense contractors spend a tremendous amount of time and monetary resources, including human capital, training courses, and documentation. (During my research, I was provided confidential access to some proprietary methods and documents so that I could analyze their internal investments and programs supporting government contracting efforts.) These large contractors know that their knowledge of how to navigate the contracting process is itself a source of competitive advantage and it is the reason they keep this information confidential. For example, Big Five contractors have created documentation they use as guidelines to negotiating FAR and DFARS clauses. One such guideline groups FAR and DFARS clauses by category and for each category describes the policy context for the clauses, outlines what the defense contractor's negotiating position should be, defines key terms, provides references to statutes, and describes the associated risks. Such guidelines are updated periodically to incorporate changes in FAR and DFARS clauses. They help contracting staff negotiate with U.S. government contracting officers and learn which techniques prove successful.

To understand the investment by Big Five contractors in terms of human capital and training courses to navigate defense R&D, let's take the example of IP. In terms of legal assistance, large defense contractors typically have more than two dozen in-house personnel to deal with IP issues and contracting, in addition to third-party law firms.

Then there is training. The sheer amount of training these defense contractors offer internally on negotiating IP in government contracts speaks to the burdensome complexity. One contractor offers nearly a dozen such courses, at considerable expense. The time employees spend in training does not contribute directly to revenue. What is at stake is clear from the following excerpt from a training course description that explains its purpose:

> A majority of [defense contractor]'s annual revenue results from contracts with the Department of Defense (DOD). *Failure to understand or comply with the particular (and often complex) rules that govern the development and retention of intellectual property rights under contracts with DOD can have dire consequences* including:
>
> - Forfeiture of critical IP rights and loss of control of [defense contractor–] created technology,
> - Loss of market differentiating technology; loss of competitive bid and competitive advantage, and
> - Loss of commercial licensing opportunities
>
> Put another way, our ability to master the concepts described in this course across all [defense contractor departments]—from program to proposal managers to supply chain and contracts—is a core capability that sustains our competitive advantage and *underpins our success as a defense contractor.*[20]

In essence the bureaucracy surrounding contract negotiations for the U.S. government is designed for large contracts and defense contractors. Training courses, internal legal staff, and reliance on broad support from outside counsel cost a lot of money and show a significant investment in the contracting process. A company would not make such an investment unless it was necessary and felt it would get a return on that investment. Large, established players have the resources available to make the investment up front. Start-ups and small businesses are often not able to do so, leaving them unable to participate or inadequately prepared to do so.

Small Businesses and Start-Ups Face Greater Challenges

The complexity of the bureaucracy leads to the additional consequence that small businesses and start-ups face greater challenges penetrating the regulatory system and reaching contract completion. As cumbersome and expensive a process it is for large defense contractors with years of experience to negotiate a defense contract with the government, it is even more challenging for start-ups and small businesses that may lack the experience and resources to navigate the regulatory regime, including specialized knowledge of the FAR and DFARS clauses involved.[21]

The importance of the resources required to engage with the government on defense contracts is underscored when we contrast large defense contractors with start-ups and small businesses trying to protect their IP while working with the U.S. government. The government's rules and regulations are set up for dealing with large companies. The IP negotiation is embedded into the contract template in the form of default provisions that are designed for very large contracts. In fact the template is identical for a contract of less than $1 million and a multi-decades' contract exceeding $1 billion. Thus even the small business contract has complex features designed to accommodate the potential complexity of major multi-year contracts but that may not be relevant to a small business's short-term project. The bureaucracy associated with the negotiation is thus a deterrent to smaller potential contractors.

In addition to my research across start-ups and small businesses, I conducted a multi-year case study of a start-up to understand the particular significance of the contracting bureaucracy and how it affects the ultimate decision of a start-up about whether to participate in secure U.S. government R&D at all. The story was, unfortunately, not unique among the start-ups I interviewed and studied.

Over the course of my five-year case study, I was able to draw on confidential data to which I was given access, and I shadowed the principals of the start-up as they went from ideation through contract. This offered a firsthand perspective on the start-up's relationship with the U.S. Air Force to conduct secure R&D. Despite recognition from two USAF senior leaders that the start-up's technology met an "urgent" and "critical" need in the USAF, the government was not able to kick off its contract with the start-up for three years due to a series of bureaucratic steps, government IT issues, and contract processing delays. For example, the DOD had to hire a web developer to resolve IT issues when the start-up submitted its application due to incompatibility between agency and DOD systems. At another point the review process was delayed several weeks due to confusion regarding why the start-up's Commercial and Government Entity (CAGE) code was not in the DOD award system. The issue was apparent unfamiliarity with CAGE code processes for companies that had not previously contracted with the DOD. Even after the award, the start-up's work was hindered by government delays in processing payments due to a software bug in the DOD's accounting system. Identifying the bug required over thirty hours of time by the start-up's employees. All this put further pressure on the start-up's limited financial and human resources.

The relationship between the start-up and the USAF began through networking at a conference; it paid off when a USAF program chief engineer determined that the start-up's technology met an urgent and critical national security need. The start-up had already invested in internal R&D to develop its technology. The start-up and USAF wanted to enter into a contract for additional R&D to develop the technology for delivery to the USAF to meet the critical security need. But even after reaching an agreement, the contracting process itself took nearly three years to complete because of the incredible bureaucracy and complexity of the regulatory regime. The start-up did not receive its first payment until more than three years after the initial government decision that the project was urgent and should be undertaken. (It turns out that such a timeline is not unusual.)

This timeline reveals just how out of sync government contracting is with the commercial sector. A three-year-plus contracting process for an *urgent* project is not only infeasible commercially, but is also at odds with the pace of business for start-ups, some of which go from technology inception to exit in that same amount of time. A three-year delay clearly hurt this start-up, which suffered financial setbacks that could have forced it to shut down. Nor is the delay helpful to the government, which needs critical technology quickly. If the government wants access to valuable innovations emerging from such companies, it must accelerate its contracting process.

In this example some delays were caused by restructuring of the USAF acquisition agency; others stemmed from incompatibility between agency websites (the "urgent" designation of this effort resulted in accelerated hiring of a web developer by the DOD to resolve the issue); and still others were rooted in a general lack of familiarity within the government contracting team in working with companies that were not already defense contractors.

At one point the lead government contracting officer refused to continue discussions upon learning that the start-up had not retained a lawyer for the contract negotiations. The rationale was that the start-up could not possibly understand the contractual machinery and specialized legal terminology. While there is no requirement for a start-up to obtain legal counsel, apparently the government officer felt uncomfortable proceeding without one. The start-up had to bring in a lawyer for the remainder of the contract negotiations. Retaining legal counsel for contracting can be demanding for a start-up that may not yet have a steady source of income. In the present example a small team was required to make an incredible investment of financial resources and time, paying for an attorney and dedicating hundreds of hours to understanding and executing the negotiation process. In fact the National Defense Industrial Association (NDIA) Small Business Division Subcommittee found that "small businesses cannot afford to enforce IP protection with either the primes or the government."[22]

While the start-up case study discussed above is of a single start-up, it is clear from my dozens of interviews with other start-up founders and government agency program managers that the story is not unique. Similar complexity and bureaucratic challenges appear to be the norm, not the exception. Most companies that start the process do not win a contract award in the end, and many start-ups change business direction before completing the process.

The case study above is also important because the start-up selected is the type of company the secure U.S. government R&D environment *wants*—i.e., a company focused on delivery of national security technology and one that did not take unfair advantage of the SBIR program (meaning that it did not rely on the SBIR program as its sole commercialization strategy). Instead this start-up was leveraging the SBIR contract to gain additional outside investment and establish a sustainable business model—the very goal of the SBIR program.

The bottom line is that the complex set of regulations that govern secure R&D make the contracting process a significant hurdle not only for established defense contractors, but especially for small businesses and start-ups, and it ends up pitting innovation against other values. An unintended consequence is that new companies, even if they have high innovation potential, are often deterred from entering the playing field or are excluded altogether, thus potentially reducing the availability and speed of innovation related to national security.

It should be noted that as this book was reaching press, the USAF hosted an inaugural Pitch Day event, achieving the incredible feat of streamlining FAR and DFARS into a single page contract and processing payment to small businesses on the same day via government credit card. Although these were unclassified contracts, it was a significant step forward in addressing the challenges of the contracting process. USAF Pitch Day builds on the important work of the USAF Commercialization Readiness Program (CRP), led by James Sweeney III, which accelerates the transition of technologies into military applications by helping connect small businesses with military customers.

Before we dive deeper into the specifics of the secrecy and participation challenges, it is worth further exploring the ways in which the regulatory regime governing secure U.S. government R&D may unintentionally favor established players over new entrants. Three features of the regulatory regime's policies in particular contribute to favoring established players: (1) entrenched policies and relationships; (2) the requirement to work with established prime contractors; and (3) incentives that have led to so-called "SBIR shops" in lieu of the commercialization of innovative technology. Let's look more closely at each of these features to determine how they became entrenched and what impact they have on full and open competition.

Entrenched Policies and Relationships

FANG, the classified U.S. government program discussed in chapter 3, represents an important effort by the government to bring open innovation to bear on critical national security technology needs. FANG tried to open the participant pool of innovators. Yet the evolution of the FANG program, starting with its predecessor programs in the 1990s, shows how entrenched government mind-sets are in valuing experience with government contracting over involving others. The result creates a tension with a key tenet of *open* technology innovation strategies—i.e., involving new participants.

The FANG program (to develop an amphibious vehicle for the U.S. Marines) has gone through more than one name change over the years, including Advanced Amphibious Assault Vehicle (AAAV) and Expeditionary Fighting Vehicle (EFV). Program leadership implemented policies in the 1990s that favored established defense contractors, the effects of which can be seen today. The story of the creation of the predecessor programs to FANG and the companies that first competed and bid on the contract is illuminating.

The AAAV program initiated a concept exploration phase in 1990. While this phase was procured competitively, the two awardees were large, established defense contractors that were also the two largest makers of armored vehicles: BAE and General Dynamics. BAE had also been the prime contractor on an even earlier predecessor program, the Amphibious Assault Vehicle (AAV).[23] No smaller players were included in the concept exploration phase of the program in 1990, perhaps a reflection of the predisposition toward established defense contractors.

At the end of the concept exploration phase, the Marines selected one company because the two resulting designs were too similar, a situation that in turn meant the choice had to be made on some basis other than the technical merits of the ideas. So the winner was selected based on three criteria that buttressed traditional contracting paradigms: the willingness of the contractor to co-locate an R&D facility and key personnel (engineers from General Dynamics had to move to Washington DC to be in the same location as each other and the government customer); performance on previous government contracts; and the willingness of the contractor to share in the cost of the development.[24] These criteria, though, did not just affect a contracting decision in the 1990s. Establishing a co-located and cost-shared R&D facility between the government and a single established defense contractor entrenched these contracting decisions for years.

After the initial competition in 1990, which ended when the government selected one participant at a very early stage ("early" even by its own typical standards), this program never had another full and open competition. General Dynamics continued as the sole participant in the concept development/validation (CD/V) phase, and, until the program's cancellation in 2011, no further contracts were procured competitively.[25]

Figure 17 is a snapshot of a 2007 contract awarded to General Dynamics to develop an alternative drivetrain for EFV (renamed from AAAV). It gives no further justification than "Unique Source" for not having "full and open competition," thus alluding to FAR Subpart 6.302–1: "(A) Demonstrates a unique and innovative concept, or, demonstrates a unique capability of the source to provide the particular research services proposed." It is not surprising that General Dynamics was in a unique position to offer a solution given that the prior award provided it unique, on-site access to the customer. The remainder of the form appears to have been filled out using drop-down menus. It does reveal a choice to limit participation: "Not Competed." The EFV program was canceled in 2011 after $3 billion in developmental funding had been spent, yet the Marines' need was still unmet.[26]

General Dynamics' CD/V award is indicative of a mind-set that values experience in U.S. government contracting over openness. While

Competition Information	
Extent Competed For Referenced IDV:	
Extent Competed:	Not Competed
Solicitation Procedures:	Select One
Type Of Set Aside:	No set aside used
Evaluated Preference:	Select One
SBIR/STTR:	Select One
Statutory Exception To Fair Opportunity:	Select One
Other Than Full And Open Competition:	Unique Source
Local Area Set Aside:	Select One
Number Of Offers Received:	1
Pre Award FBO Synopsis:	✓
Small Business Competitiveness Demonstration Program:	
SBA/OFPP Synopsis Waiver Pilot:	
Commercial Item Test Program:	
Alternative Advertising:	
Commercial Item Acquisition Procedures:	
A76 Action:	

Preference Programs / Other Data	
Contracting Officer's Business Size Selection:	Other than Small Business
Subcontract Plan:	Plan Required - Incentive Not Included
Price Evaluation Percent Difference:	0 %
Reason Not Awarded To Small Disadvantaged Business:	No Known SDB Source
Reason Not Awarded To Small Business:	No Known SB Source

FIG. 17. M67854-01-C-0001 contract to develop an alternative drivetrain for EFV. Source: Snapshot from the Federal Procurement Data System (General Services Administration, "Award M6785401C0001").

the early move to select a single contractor is understandable, the selection in the 1990s likely had long-term consequences on competition and innovation for the government's amphibious vehicle needs.

The unmet government need related to amphibious vehicle technology resulted in the FANG program—a renewed attempt to find a solution. When FANG was canceled in 2013, the U.S. Marines restarted the work with General Dynamics; the work has since been renamed the Amphibious Combat Vehicle (ACV). General Dynamics was awarded an ACV contract in June 2014.[27] Possibly in recognition of the risks of depending on a single vendor to solve critical technology problems, the Marines reopened the playing field by introducing new vendors to compete for the ACV contract, all of which were traditional defense contractors.[28] In 2018 the contract to built the new ACV was awarded to BAE, the defense contractor who built the original AAV.[29]

Outside the amphibious vehicle story there are other examples of entrenched policies and relationships that favor established players. Contracting mechanisms or "contract vehicles" that streamline the acquisition process can sometimes disadvantage small businesses. For example, the so-called "omnibus contract"—also referred to as "bundling"—consolidates several requirements or procurements into a single contract to reduce administrative costs and time. Contract vehicles of this sort, however, "decrease[] opportunities for small business participation."[30]

Another example is "indefinite delivery, indefinite quantity" (IDIQ) contracts for services in which the U.S. government specifies a contract period (usually a few years) but does not need to define specifically the cost or all requirements at the time of the contract initiation. The IDIQ contract vehicle is often praised for the flexibility it offers; once an IDIQ contract is in place, the government can issue task orders for specific services against basic requirements. Investigations have shown, however, that some IDIQ contracts discourage small businesses from competing.[31]

Even when DARPA implemented a program designed to encourage individuals and small businesses to participate—the CFT program (introduced in chapter 1)—there were problems with the IDIQ contract. DARPA funded a prime defense contractor with an IDIQ contract, and that contractor in turn made sub-awards to CFT participants. But the CFT solicitation stated that "Selectees that take major exceptions to the agreement's terms and conditions may experience a longer timeframe before the subcontract is awarded or may not receive an award."[32] This is an example of the second of the three particular features of the regulatory regime's policies that favor established players: the requirement that businesses work with an established prime contractor.

Requirement to Work with an Established Prime Contractor

Since established prime defense contractors are often supported by entrenched policies, it may be a strategically advantageous *choice* for small businesses to partner with these primes to deliver technology to the U.S. government. In many cases, however, such a partnership is practically a *requirement*. My research with small businesses reveals that government agencies are unlikely to award certain contracts to a small business unless it is partnered with a prime.

Take the example of one small business I studied whose SBIR proposal to the DOD was rejected. Confidential company materials revealed that feedback regarding the technology and the commercialization plan from the SBIR program manager was positive.

The program manager even recommended resubmission, but only after a correction was made in one aspect of the proposal: "The lone element missing from the commercialization plan is the absence of [a] letter of endorsement or interest from a major military entity, certified government prime."[33]

The implications of this "requirement" are significant and may have unforeseen consequences. Seeking out a relationship with a prime defense contractor not only adds another burden on a small business, but such a relationship may also create risks for small businesses. For instance, the prime contractor may attempt to co-opt financial incentives and IP from the small business; it may impose requirements additional to those already imposed by regulation as a precondition to engagement; and it may leverage its position to obtain favorable (but likely asymmetric) terms and conditions in an agreement with the small business.[34] For better or worse, such a system deprives the small business of autonomy.

As a further consequence, the forced involvement of a prime defense contractor may impede the flow of innovation to the government. The needs of the government and the small business may be aligned, but introducing a third party (the prime defense contractor) with a third set of needs that may not be aligned can derail or significantly delay the process, as figure 18 shows. Consider a situation where the small business is developing innovative technology supporting national security that the government wants. The prime contractor is typically involved to provide some sort of risk mitigation because of its prior experience working with the government. But while the experienced contractor can be quite valuable given the complexities of the U.S. government bureaucracy, the system-level impacts of effectively requiring the prime defense contractor as an intermediary may not be fully understood by the U.S. government officials who impose these requirements implicitly or explicitly.

For example, there are situations where imposing an intermediate ends up excluding certain groups of innovators that are of value to the U.S. government. In one case, a venture-backed start-up had technology with both defense and commercial applications. The prime defense contractor initiated the start-up's exploration of its defense applications. But there was a wrinkle: the venture capital-

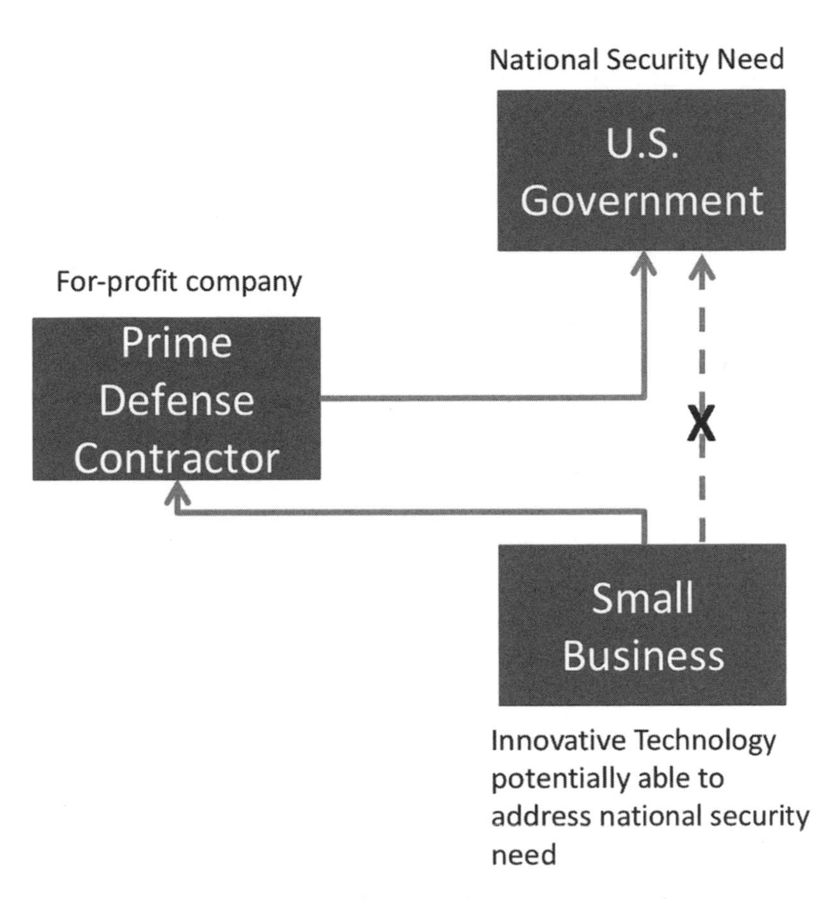

Fig. 18. Potential misalignment of needs and incentives among the government, prime defense contractors, and small businesses. Created by the author.

ists on the start-up's board of directors, crucial to the business's continuation, insisted that the market size or specific application be defined. The proprietary and competition-sensitive nature of the prime defense contractor's information hindered the ability to do so. Thus the lack of information sharing between the small business and the prime contractor affected the small business's internal ability to justify the project.

In another case a small business experienced in government contracting was working with a prime defense contractor on a national security–focused R&D project. The two contractors debated nearly every term and condition during the initial contracting process, protracting the process but ultimately leading to an agreement. After work was initiated, key individuals from the small business spun out a new start-up. However, the new start-up was unable to continue the relationship with the prime defense contractor because of the prime's metrics of the supplier risk profile. The new start-up had no track record of performance. Thus it was not the government's requirement that directly precluded the new start-up's participation but rather the prime contractor's own requirement. Whether the prime's own requirements were justified or not, it warrants reflection when a private intermediary, rather than the government, controls participation in the innovation process.

In a third case multiple small businesses were seeking to partner with a prime in the development of a new capability. The small business that was selected was a start-up that had spun out of university research and was staffed by students and professors. Confidential correspondence within the prime's organization revealed that the prime preferred that particular start-up because it would agree to any terms and conditions the prime stipulated. It negotiated nothing. Here again the private intermediary controls participation in the innovation process, and some of its reasons for selecting participants may be tangential to who is best suited to solve the government's national security need.

These examples show that certain groups of innovators may be unintentionally excluded from the process of innovation in areas of importance to national security. Whether it is because they lack a track record of performance, have investors who require certain information, or show a propensity to negotiate prime defense contractors' standard terms and conditions for partnership, these groups are being excluded because of a potentially widespread "requirement" to partner with a prime defense contractor—in a context in which the U.S. government has little control over the relationship between these innovators and the primes.

Related to the concept of a requirement to work with a prime is the Lead System Integrator (LSI) concept, wherein the U.S. government hires a prime to oversee and execute a complex defense acquisition program. This concept gained popularity in the mid-nineties but was subsequently found to have many of the same drawbacks as the previous example, such as lack of transparency on selecting subcontractors.[35] Legislation was enacted to move the U.S. government away from the LSI concept, yet relatively little has been done to address the "requirement" for small businesses to partner with primes and its potential detrimental impact on national security innovations reaching the U.S. government.[36]

SBIR Shops

Another unintended consequence of the regulatory regime is the proliferation of so-called "SBIR shops"—also known as "SBIR houses" or "SBIR mills"—which have built a cottage industry of serially applying to and completing SBIR grants without ever fully commercializing their technology. These companies are established players well-versed in contracting and winning grants, but they are believed to do little to actually advance long-term innovation. Government agencies and commercial players alike tend to view SBIR shops in a negative light. As one venture capitalist put it, SBIR shops "game the system" by "mov[ing] from agency to agency."[37] They are "like leeches feeding off these government programs without any real interest . . . in bringing their technology into the real world. . . . They are hurting innovation in the long run" by "sucking away precious resources from the system."[38]

As discussed in chapter 2, eleven federal agencies reserve a portion of their R&D funds for small businesses through the SBIR program, which totals approximately $2 billion annually.[39] These agencies post descriptions of their problems/needs, and small businesses submit proposals with their technical ideas to address the needs and their plans for commercialization of the technology. SBIR projects occur in three distinct phases. Phase I projects are about $200,000, one year in duration, and focus on demonstrating the technical feasibility of a proposed technology. Phase II projects are usually $1 million, two years in duration, and represent a continuation and expansion of the R&D. Phase III is unique; there is no specified duration or amount. "SBIR Phase III refers to work that derives from, extends, or logically concludes effort(s) performed under prior SBIR funding agreements, *but is funded by sources other than the SBIR Program*."[40]

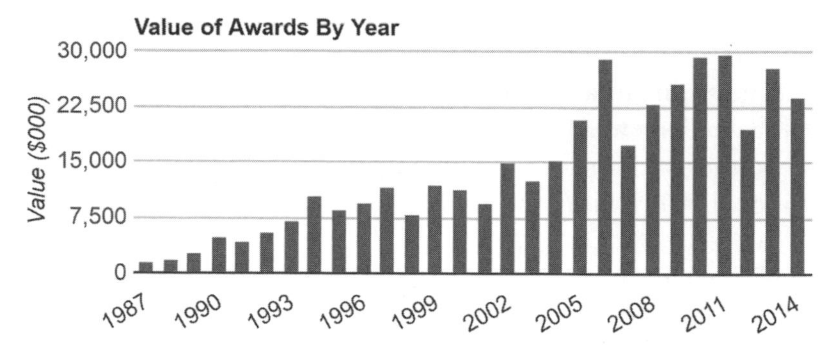

FIG. 19. SBIR awards over time for one anonymous SBIR shop. Source: SBIR source.

A review of data spanning twelve years of the SBIR program confirms the reality of SBIR shops. Eleven companies have received in excess of five hundred SBIR awards each. For example, one company (left anonymous here) has won more than seventeen hundred awards.[41] Characteristic of an SBIR shop, the firm's awards span ten of the eleven agencies participating in the SBIR program, including a dozen different divisions with DOD. Figure 19 breaks down the value of these awards over time.

It is possible that the regulatory regime unintentionally encourages "SBIR shop" behavior. The SBIR program's "Company Commercialization Report" tracks a firm's previous SBIR Phase II projects, and such tracking is required in every SBIR proposal. For every potential SBIR award, a firm is judged based on the presence of a previous Phase II project and on revenue obtained for the Phase II project other than through the SBIR program. All eleven federal agencies that participate in the SBIR program agree to a commercialization benchmark requirement, which currently states that the SBIR applicant must have received at least $100,000 in sales and investment beyond the SBIR Phase II award.[42] To put that in context, $100,000 is less than 10 percent of the combined amount of SBIR Phase I and Phase II awards. A company applying for an SBIR award needs only to exceed a 10 percent commercial interest, compared to the government investment, to be judged favorably against the evaluation criteria for potential for commercialization. Although the objective is to reward the commercialization of previous ideas, the DOD's 10 percent standard for successful commercialization may be unintentionally incentivizing firms to obtain a large percentage of their revenue through SBIR monies (like the firm in fig. 19) while never taking an idea fully down the commercialization path.

TABLE 3. Template company commercialization report
Full report and company certification

Commercialization achievement index	N/A
Phase I Awards	0
Phase II Awards	0
Number of patents resulting from SBIR/STTR	0
Firm's total revenue	<$100,000
SBIR/STTR funding as percent of revenue	0%
Current number of employees	0
Year founded	2019
IPO resulting from SBIR/STTR	No

As these analyses and stories show, the burden imposed by complex regulatory regimes is more than a mere inconvenience to potential innovators. Indeed the complex regulations that were developed to ensure fairness, rule of law, and other constitutional principles have also had unintended consequences that deter and in some cases exclude new entrants from participating in the innovation process. Policies that were developed over time to serve one value have in some cases resulted in unintended consequences in other areas of the overall system that obstruct the competing goal of bringing forth innovation and maximizing participation from potentially innovative companies.

FIVE

Secrecy versus Open Innovation

The central mission of secure U.S. government R&D is national security. Indeed, as discussed in chapter 4, the national security objective is enshrined in the mission statements of the agencies that oversee secure government R&D. At the same time, laws and policies designed to support national security objectives that comprise the regulatory regimes governing such R&D sometimes create tensions with the competing interest of innovation—despite that national security and innovation are not inherently contradictory objectives.

Many policymakers and enforcers of policies mistakenly conflate national security and secrecy because of an underlying assumption that they go hand in hand—an assumption even the U.S. government has made: "Our Nation's progress depends on the free flow of information both within the Government and to the American people. Nevertheless, throughout our history, the national defense has required that certain information be maintained in confidence . . . to protect our citizens, our democratic institutions, our homeland security, and our interactions with foreign nations."[1]

Even the definition of "top secret" comes with the assumption that national security requires utmost secrecy: "'Top Secret' shall be applied to information, the unauthorized disclosure of which reasonably could be expected to cause exceptionally grave damage to the national security that the original classification authority is able to identify or describe."[2] Further, in its "Security Provisions" section, the NSF Act of 1950 asserts that national security *requires* the utmost secrecy: "Matters relating to the national defense . . . shall establish such security requirements and safeguards, including restrictions with respect to access to information and property, as [the relevant official] deems necessary."[3]

In reality national security and secrecy—though related—are different. Secrecy often furthers the interest of national security, but other times it does not. In some instances secrecy may be a detriment to national security. For example, solving a technical problem is difficult when the specific details of the problem are obscured from potential innovators for national security reasons. Conflating secrecy and national security erects walls to innovation that ultimately hurt national security—creating the "secrecy challenge" referred to in this book.

To understand the secrecy challenge, it is useful to explore briefly why secrecy is often used as a mechanism to support national security. Where does secrecy serve a national security interest? One area is where secrecy surrounds the development of a new capability that would create strategic surprise for the United States. For example, the government kept secret the Manhattan Project in an attempt to enable the United States to develop atomic weapons capabilities faster than its adversaries. The world knew that there was competition between the United States and Germany to reach the capability first, but details of how to build an atomic bomb (such as refining uranium or arranging the fissile material into a critical mass) were kept secret to give the United States a time advantage over Germany.

Secrecy can also further national security when it is used to protect information about specific vulnerabilities, such as weaknesses in a material, code flaws, or bandwidth limitations of a radar system. For example, after bombing Hiroshima and Nagasaki, the United States required lead time before it could deploy a third bomb and significant lead time before it could build more bombs—facts kept secret at the time so that adversaries did not know the United States was not prepared to immediately strike again. In the case of FANG, discussed in chapter 3, some requirements are classified, such as those regarding how rugged the vehicle's armor must be for survivability, because their disclosure could be exploited by an adversary.

A related area is the protection of small pieces of information that, when put together, would reveal a larger strategic picture—referred to in secure R&D environments as operations security (OPSEC). For example, an adversary who had supply-chain information for key components of a technical system could exploit it by targeting specific manufacturing links or factories. Information about movements of the specific aircraft carrier used to transport the atomic bombs in the Manhattan Project could have been similarly exploited if not kept secret.

While secrecy limits the ability of adversaries to exploit system vulnerabilities, there is an argument that revealing selected information about desired capabilities or vulnerabilities enables innovation to solve and address these issues. For example, potential innovators cannot work to address material weaknesses in an aircraft's structure unless they know what thc material is and that there is a weakness in the first place.

A rigid commitment to secrecy often precludes full and open competition, as well as information sharing, two factors important to the very innovation desired to serve national security interests by advancing American technological superiority. The commitment to secrecy without a rigorous consideration about whether that secrecy truly serves the underlying national security interest can, therefore, be in tension with innovation because secrecy limits participation and creates silos within secure government R&D. The FANG story, as well as stories from contractors and businesses in the field, reinforces the reality and persistence of this tension.

By definition secrecy regulations prevent the free flow of information and open participation, in turn inhibiting innovation—as many researchers have explained. The free flow of information and open information provide "access to a wide array of ideas and knowledge," which leads to "discovering radical new solutions to solving problems."[4] A potential innovator "can benefit from a larger solver population because he obtains a more diverse set of solutions."[5] Accessing a "larger pool of skilled contributors" has benefits.[6]

DARPA Program Tenure

The "silo" effect and its impact on accessing ideas from a broad range of likely innovators is a persistent systemic challenge. This is seen even in DARPA, a government agency recognized for its cross-cutting innovation. An analysis of the innovation process across several DARPA programs illustrates the challenge, which is caused by a combination of the program manager (PM) tenure structure and the pervasive emphasis on secrecy.

DARPA imposes four- to five-year limits on the tenure of its PMs to create a sense of urgency and help with recruiting top talent.[7] While this policy has had success in jump-starting certain innovation, in other ways it can hinder innovation when implemented alongside strict secrecy measures. First, the short tenure requires an aggressive and deliberate approach to establishing new secure programs. Within the first year or two a new project manager at DARPA must juggle the demanding time requirements of continuing and wrapping up the prior PM's programs and identifying an end user or customer need area aligned with the PM's own background and knowledge. The PM then has to establish the new program, "selling" it internally to gain approval. By the time approval is granted, PMs may be concerned about whether their few remaining years at DARPA are enough to successfully conduct the R&D required to innovate and achieve the promised technological advancement and meet the national security need.

Second, secrecy issues begin to affect the innovation process once approval for a new program is granted by limiting the potential innovators included in the proposal and ideation process. This limitation arises because many DARPA programs are for "special access" secure R&D, meaning they are classified, and only those individuals (not companies) with a "need to know" and an appropriate security clearance are informed of the problem and need for a solution. Secrecy restrictions prevent openly posting the specific need to a public forum, so a PM must identify potential innovators in some other way. The PM may invite companies to a classified "Industry Day" but then PMs must rely on their own personal network to strategically select those with a "need to know" and the ability to contribute an innovation to the specific need area.

Third, DARPA project managers often provide a very short period after an Industry Day for initial ideas to be proposed before the evaluation and contracting stage. They expect ideas to improve over the course of the R&D contract. This process occurs with the limited set of innovators that have been afforded the need-to-know—and who may have been selected based on a personal connection rather than a more open, merit-based process that would be possible without the secrecy limitations.

On the surface it may seem that a quick solution to this participation challenge for innovation would be to increase the tenure of PMs working on secure programs to relieve them of the time pressure of establishing their programs, thereby enabling them to be more thoughtful about using strategies like open innovation. However, it is not so simple. Such an approach would also deprive secure programs of the other benefits of short tenure (such as a sense of urgency and access to top talent). More beneficial, then, would be to adopt technology innovation strategies and support regulatory regimes that maintain these benefits while also opening up the innovation process without compromising security.

It is important to note that the tension with innovation described here derives from secrecy, not national security itself. Thus were it possible to reduce or work around secrecy requirements and thereby enhance innovation that contributes to technological superiority, it would actually advance national security interests rather than hurt them. Some government agencies are already trying to address this issue, but their authority to implement change is limited, and more concerted action from the executive or legislative branches will be required for meaningful change to occur.

Current Regulatory Regime Approaches

Current regulatory regimes employ several approaches as they attempt to address the tension between innovation and secrecy. The first is a recognition that classifying information limits innovation. So the approach is to avoid classifying information that does not need to be classified (i.e., avoid overclassifying) and, where possible, selectively declassify information.

The Bush administration executive order on "Classified National Security Information" was modified by the Obama administration in 2009 to emphasize this approach. As President Obama specified, "If there is significant doubt about the appropriate level of classification, it shall be classified at the lower level," and "No information may remain classified indefinitely."[8]

Chapter 1 showed that open technology innovation strategies can accelerate the innovation process; enabling a broader innovator community to participate results in the rapid generation of a diversity of ideas. If aspects of a classified problem can be selectively declassified to leverage open technology innovation, it may allow a greater number of innovators to address an urgent national security need more quickly than when classification closes it off to a more limited group of innovators. Even though this approach could have significant effects on enabling technology innovation, it remains impossible in many cases because current regulatory frameworks do not empower program managers to make such tradeoffs.

It is a catch-22. The U.S. government wants to focus technology innovation on specific vulnerabilities of U.S. systems and assets or in areas where U.S. technological superiority is threatened. These vulnerabilities, though, are classified; disclosing them could enable an adversary to exploit them to attack the United States. Thus it is difficult or impossible to leverage open technology innovation strategies in some situations. As Chesbrough has described, it is difficult to solicit ideas if a problem cannot be fully defined.[9] The question the government must answer, then, is what can it reveal? If a government agency believes open technology innovation can advance the state of the art and provide a technology solution that helps address a critical vulnerability, it would need to declassify at least selected aspects of the vulnerability—that is, obscure the full vulnerability but reveal some of the technology requirements. That tradeoff calculation would need to be made based on whether potential innovation is more valuable to national security than keeping every aspect of the vulnerability classified.

The authority that government agencies have to make this tradeoff is, in many cases, limited. Only original classification authorities (OCAs) are able to classify or declassify information (see chapter 2), but even OCAs lack the power to make such tradeoffs because they must follow policy guidance. Current policy guidance specifies that any information that could cause serious damage to national security must be protected at the "secret" level, even if it is possible that greater damage would be caused by keeping information classified and restricted to closed innovation strategies.

One way to deal with this problem would be to empower government agencies with decision-making authority to make the tradeoff between the national security value of keeping information classified and the national security value of potentially accelerating the time to addressing a vulnerability by leveraging open innovation. Based on a systems architecture approach, decision makers could use specially developed tools to identify and isolate a subset of the system architecture to declassify for the purposes of leveraging open technology innovation while minimizing the exposure of vulnerabilities that an adversary might exploit.

The second approach is simply to have a classified version of open innovation platforms that is available to innovators with security clearances, thus ensuring that at least this innovator community can access collective knowledge and share ideas to collaborate. This approach is broader than traditional DOD R&D because secure programs tend to lead to silos, as discussed above.

There are at least two ways regulatory regimes have explored this approach. One is through the launch of the DODTechipedia website and later the NeedipeDIA website. DODTechipedia, first launched in 2008, allows two-way exchange of needs and innovations. Part of the Obama administration's Open Government Initiative, it is structured as a wiki that helps "in finding internal and external technology innovations that address the Department's capability needs."[10] Whereas a member of the public using NeedipeDIA who has an idea to address a posted need must propose a solution by following the traditional government R&D contracting and acquisition process, DODTechipedia users can post information about a solution directly to the website (with no specific monetary incentive to do so).

The Open Government Initiative describes the benefits this way: "DODTechipedia levels the playing field to ensure that everyone with good ideas, even if they have never worked with the Department of Defense, can share them with decision makers. In addition, DODTechipedia ensures that Combatant Commands have access to the information they need from the science and technology community to make the best investment decisions."[11] There is also a classified version of DODTechipedia.

While there has been no independent study of its effectiveness, DODTechipedia does make its own claim of success, indicating in 2014 that it "has over 6400 registrants who have logged over 150,000 page views while creating 30 separate blogs and more than 900 technology area, interest area and organization pages. DefenseSolutions.gov has received 37 innovative ideas on its initial theme area, six of which are now being considered for funding."[12] There is no readily available information regarding the classified version specifically.

The DIA launched its NeedipeDIA website in December 2013. Before that the traditional method of soliciting ideas to meet needs was to collect and approve needs for release in a process so lengthy that when the lists were finally released, many needs could potentially be outdated. Now on NeedipeDIA, DIA end users can directly post their needs publicly. Government R&D requests for proposals (RFPS) can reference NeedipeDIA directly, so the latest needs are always listed. DIA itself notes that NeedipeDIA is different from other processes, calling it a "rapid and agile process that is responsive to evolving mission needs."[13] Of course publicly listed needs are unclassified. In 2014 DIA launched a classified version of NeedipeDIA to address classified needs and enable the community of people with security clearances to participate better in the innovation process. At this writing, it is too early to determine the effectiveness of the classified version of NeedipeDIA. Yet even the classified version of NeedipeDIA does not address the growing number of SAPS. In a 2010 article in the *Washington Post* Dana Priest and William Arkin described just how widespread SAPS are in the secure U.S. government R&D environments:

> Beyond redundancy, secrecy within the intelligence world hampers effectiveness in other ways, say defense and intelligence officers. For the Defense Department, the root of this problem goes back to an ultra-secret group of programs for which access is extremely limited and monitored by specially trained security officers.

> These are called Special Access Programs—or SAPS—and the Pentagon's list of code names for them runs 300 pages. The intelligence community has hundreds more of its own, and those hundreds have thousands of sub-programs with their own limits on the number of people authorized to know anything about them. All this means that very few people have a complete sense of what's going on.
>
> "There's only one entity in the entire universe that has visibility on all SAPS—that's God," said James R. Clapper, Undersecretary of Defense for Intelligence.[14]

SAPS would be unable to use the approach of NeedipeDIA or DOD-Techipedia unless selective declassification, or at least selective reduction of classification from the SAP restriction, was enabled.

A third approach to addressing the tensions with national security from a secrecy perspective is to not reduce competition unless necessary. The regulatory regimes have supported full and open competition by requiring completion of a justification form if competition is reduced. Subpart 6.3 of the Federal Acquisition Regulation spells out seven reasons or "exceptions" an agency can use to justify not having full and open competition on a defense contract; one of them is "national security."

The notes on the legislative history for this law are long and confusing to a layperson. In general it is understood that competition should not be reduced unless necessary. As one researcher noted, "Whether or not an exception is fully justified, exercising these exceptions always presents a risk because without competition there is not a direct market mechanism for setting a contract price and there is no assurance that the most innovative technology available was identified."[15] Further, the exception justification forms are not always completed properly. As the GAO reported, "Justifications provided limited insight into the reasons for the noncompetitive award or did not fully describe actions that the agency could take to increase future competition."[16] There is evidence that justification form may not effectively deter reduced competition. In a March 2017 report the GAO called for action to increase competition, noting that only 55.4 percent of defense contracts were competitive in 2015, down from 58.4 percent in 2011.[17]

Most often agencies requested the exception because only one responsible contractor can satisfy the requirements—the first in the list of seven justifications allowed. (However, as discussed in chapter 6, there may be an underlying lack of incentive to innovate in the government R&D context that is driving the limited contractor pool.)

The regulatory regime has tried to increase competition. A 2009 Office of Management and Budget memorandum set a goal to reduce by 10 percent the dollars obligated to high-risk contracts (including sole-source and non-competitive contracts) by 2010.[18] The GAO estimates that a reduction of less than 1 percent was achieved.[19]

The National Security "Exception" to Increasing Competition

Most of the above relates to the most common of the permissible exceptions to increasing competition: only one responsible contractor can satisfy the requirements. But what of the national security exception? A 2012 GAO report addresses this directly, again characterizing national security as going hand in hand with secrecy: "DOD's use of the national security exception is necessary in certain situations when disclosing the government's needs in a full and open competition would reveal information that would harm national security."[20]

The GAO explains that it cannot even analyze the exception fully. "Gaps in federal procurement data limit GAO's ability to determine the full extent of DOD's use.... However, DOD intelligence agencies and special access programs frequently use the exception, but are generally excluded from reporting procurement data.... DOD policy on reporting sensitive procurements for other military department programs is not clear."[21]

As for data to which the GAO did have access, it found that justification forms were filled out improperly or lacked detail: "For most national security exception contract actions GAO reviewed, DOD used a single justification and approval document that applies to multiple contracts—known as a class justification. Among those reviewed, $3.3 billion of $3.4 billion was obligated under contracts that used class justifications, which reduce the steps required to proceed with individual contract actions that do not use full and open competition."[22] In other words, class justifications were employed that made it easier to use the national security exception. Further, 84 percent of the time the national security exception was used, it was not that competition was reduced to a smaller group, but rather that there was no competition at all.[23]

Even though it may reduce competition and inhibit innovation, not everything about class justification is viewed as negative. "According to contracting officials," notes the GAO, "the increased flexibility of national security exception class justifications helps meet mission needs."[24] Agencies sometimes see the justification form as a step that slows the government down rather than as a positive tool for finding opportunities to enable competition and innovation. Indeed this mind-set was articulated again and again in the many interviews I conducted with government acquisition organizations and defense contractors. Whatever value there is in speedy contracting, the national security exception should not be used when it isn't necessary just to speed things up. A balance is important because while class justifications may serve a near-term mission need more quickly, bringing in more innovation through greater competition is critical to long-term security.

Clearly the tension between national security and secrecy inhibits innovation that is important to advancing U.S. national security. It must be addressed. While current approaches have not yet fully addressed this tension, there does appear to be a growing awareness of the potential consequences for innovation that advances national security interests. Those aspects of the regulatory regime that create this particular tension may therefore be ripe for change.

All too often secrecy is protected without any mechanism for rigorous debate regarding whether it is required for national security. That is precisely why the ultimate regulatory policy goal should be to work around or lessen secrecy *if* it can be done in a manner that furthers *both* innovation and national security. That has been the aim of the approaches discussed above. These approaches are in the early stages. We do not yet know whether they can achieve their goals within the constraints of secure government R&D environments. That makes it important to identify and understand the people and organizations with the authority to determine what level of secrecy is required for national security.

Who Decides?

Today U.S. government security classification is largely addressed by the executive branch. The president issues executive orders that define levels of classification, identify those with the authority to classify information, and set how long information should remain classified.

Prior to World War II classifying information or restricting access was unique to the military, which defined security classification authority and decisions. In 1940, during the lead-up to U.S. involvement in the war, President Franklin D. Roosevelt issued Executive Order 8381, stipulating the president's right to classify information. This order pertained specifically to information regarding military and naval installations, and it cited statutory authority, meaning the president's authority to provide himself this right derived from statute.[25]

After the war tensions with the Soviet Union began to rise almost immediately. Many historians mark the start of the Cold War as 1947, during the presidency of Harry S. Truman. In 1951 President Truman went beyond reference to statutory authority alone, citing constitutional authority for the first time in his Executive Order 10290: "by virtue of the authority vested in [the president] by the Constitution and statutes."[26] President Truman's order defined the categories of classified information ("top secret," "secret," etc.) and those with the authority to classify information and set limits on dissemination. Whereas prior executive orders had applied only to the military, President Truman's order "marks the first time that classification system was applied to non-military federal departments and agencies."[27] The order also prohibited classified information from being disseminated outside of the executive branch.

Of course, context is relevant. Executive orders, as with other policy-making mechanisms, have specific near-term objectives that are necessitated by the geopolitical climate at the time they are issued. President Truman's stated objective in Executive Order

10290 was to avoid "unnecessary delay in the handling and transmission of documents," so he introduced the policy of classifying information at the lowest level possible.[28] The geopolitical environment at the beginning of the Cold War contributed to the perceived need to limit access as well as to facilitate military operations by providing access to classified documents without delay.

Since then some presidents have maintained the policy of classifying information at the lowest level possible. For instance, President Obama issued Executive Order 13526 in 2009; it is still in effect at this writing. Its stated objective is transparency and the support of the "free flow of information both within the Government and to the American people."[29] President Obama's executive order was issued in the geopolitical climate of the protracted Iraq war and controversy over the lack of government transparency, fueled in part by WikiLeaks, an organization that began publishing leaked classified information in 2006.[30]

Presidents since President Truman have issued their own executive orders relating to classification, generally citing both constitutional and statutory authority but without any specific reference; this includes President Obama's 2009 order. One could reasonably infer that presidents are citing their so-called "commander-in-chief powers," granted by the Constitution's general statement that the president is the head of the armed forces. The Constitution grants no specific powers regarding security classification of information, nor does it grant the executive branch exclusive authority to determine whether secrecy is required for national security. So while from a legal perspective the breadth of a president's authority over classifying information is an open question, it is certainly clear that the president (and, by extension, the executive branch) is likely to be at least one of the parties with authority to determine how decisions are made regarding the necessity of secrecy for national security. While the executive branch has taken the lead, the legislature has also been involved.

Congress has enacted or debated legislation related to the dissemination of classified information. For instance, Congress passed the Atomic Energy Act in 2011 and "established a separate regime . . .

for the protection of nuclear-related 'Restricted Data.'"[31] The Government Secrecy Reform Act of 1999 had the Senate considering legislation to "provide for a system to classify information in the interests of national security and a system to declassify such information."[32] Therefore, given that there is no provision of exclusive authority by the president over classified information, it is possible for Congress to weigh in on the issue of secrecy and support the initiatives to reduce secrecy to the extent possible while backing open technology innovation in areas of importance to national security.[33] Of course it stands to reason that joint action by the executive and legislative branches would likely create the most sustainable change.

A Pattern of Reversal

One of the main challenges to introducing more open innovation in the national security context is the inconsistency of executive orders; such inconsistency makes it difficult to try out new approaches and gauge their success over time. Presidential priorities and values change from one administration to the next when a new president comes from a different political party than his predecessor, as is revealed by the pattern of reversal of executive orders shown in table 4. Each time the president's party changes, a new executive order on security classification is issued.

TABLE 4. Demonstrated pattern of reversal of security classification policy

President	Party	EO on Security Classification	Year
Harry S. Truman	D	EO 10290	1951
Dwight D. Eisenhower	R	EO 10501	1953
John F. Kennedy	D	EO 10964	1961
Lyndon B. Johnson	D	—N/A—	
Richard Nixon	R	EO 11652	1972
Gerald Ford	R	—N/A—	
Jimmy Carter	D	EO 12065	1978
Ronald Reagan	R	EO 12356	1982

George H. W. Bush	R	—N/A*—	
Bill Clinton	D	EO 12958	1995
George W. Bush	R	EO 13292	2003
Barack Obama	D	EO 13526	2009

* Modified EO 12356 but did not issue a new EO.

As an illustration of the reversal problem, consider the executive orders on security classification from three successive presidents: Bill Clinton, George W. Bush, and Barack Obama. President Clinton's Executive Order 12958 in 1995 sought to reduce secrecy where possible. It set ten-year limits on newly classified information in most cases; classifiers were instructed to keep information unclassified when in doubt or at the lowest level possible; and declassification was mandated for information twenty-five years old (with limited exceptions). In addition, the order established an Interagency Security Classification Appeals Panel (ISCAP) to handle appeals to the mandated declassifications, with appeals to ISCAP rulings made directly to the president.

President Bush's Executive Order 13292, issued in the wake of the September 11 attacks, in general lessened the amount of disclosure required by President Clinton's order. President Bush's order removed the directive to disclose information when in doubt, extended the starting date for automatic declassification, and increased the scope of types of documents that could be classified.[34] An analysis by the group Public Citizen concludes that President Bush's order tends toward greater secrecy than President Clinton's order: "Although Executive Order 13,292 retains most of the structure of the 1995 Order, it modifies many of the critical provisions in ways that encourage greater secrecy and allow agencies to postpone or avoid declassification that would have been required under the 1995 Order."[35]

President Obama's Executive Order 13526 replaced the previous two orders and generally represented a return to the intent of President Clinton's order, including the reestablishment of President Clinton's directive to disclose information when in doubt. President Obama's order reduced the number of people authorized to originate classification and established the National Declassification Center to address the ISCAP backlog and help shorten the time required to declassify documents. Further, Order 13526 mandates that agencies report annually to the Information Security Oversight Office (ISOO). Still President Obama's Order 13526 did not fully restore President Clinton's Order 12958.

The pattern of reversal affects the reclassification of declassified documents as well. President Clinton's order specifically disallowed reclassification; President Bush reversed that ruling to allow reclassification by agency heads or deputy agency heads;[36] and President Obama's order, while swinging the pendulum back toward President Clinton, still allowed reclassification in certain cases. A 1997 congressional committee report drew attention to this problem of inconsistency: "Repeated changes both disrupt the efficient administration of the classification system and can be very costly," wrote its authors, Daniel Moynihan and Larry Combest. "Each new order has required that agencies devote significant time and resources attempting to make personnel aware of how policy changes affect their work."[37]

Even just the *threat* of policy reversal can play a role. Such a threat may explain the slow enactment of President Obama's Executive Order 13526. For example, it took the Department of Homeland Security *five years* to issue a statement that it was *beginning* to update regulations to incorporate new and revised procedures pursuant to the order.[38]

The ability to block rulings by ISCAP reveals another aspect of how this inconsistency plays out in the context of secrecy versus open innovation. President Clinton's order established ISCAP to handle appeals to the mandated declassifications, with appeals to ISCAP decisions going directly to the president. President Bush's order provided unique authority to the director of central intelligence (DCI; now known as the director of national intelligence, or DNI) to block declassification rulings by ISCAP without consulting the president.[39] The DCI was given authority to reject ISCAP rulings if he or she determined that disclosure would cause damage to national security; further, "the [DCI's] decision to override

ISCAP can be reversed only by the President"[40]—meaning the DCI does not need to provide justification to another person or organization to set aside ISCAP's decision to declassify a document and keep it classified. President Obama's order swung the pendulum back toward the spirit of President Clinton's order, affording the DCI/DNI no special privileges to block ISCAP rulings. President Obama reinstated the policy that any agency head wishing to appeal an ISCAP decision must present the appeal to the president, but it added that the appeal would be made through the national security adviser.

The ISOO submits an annual public report to the president that provides aggregate data on the number of documents that are declassified, the number of appeals made to ISCAP, and the costs of declassifying information (among other information). One ISOO report reveals that President Clinton's order was "so successful in promoting declassification that, during the first six years after this Order was issued, the average number of records declassified each year increased more than tenfold."[41] But another ISOO report revealed that "during the first full fiscal year of the Bush Administration (2001), the total number of classification actions increased by 44 percent to 33,020,887."[42]

Figure 20 shows that of the slightly over fifteen hundred documents whose declassification was appealed, ISCAP proceeded with at least partial declassification of the majority of them. As figure 21 shows, most appeals to ISCAP come from the intelligence community, including the CIA, National Security Agency, National Security Council, and Defense Intelligence Agency.

An array of factors related to the ISCAP process contributes to the problem that secrecy continues to thwart open innovation. For instance, as of this writing, the CIA director has the unilateral authority to reverse an ISCAP decision to declassify documents, as per President Bush's order. Allowing the head of the same agency that submitted a classification request to ISCAP to reverse any ISCAP decision with which it disagrees would seem to defeat the very purpose of ISCAP and circumvent the policy goal of decreasing secrecy.

Further, a large number of the documents that are marked as

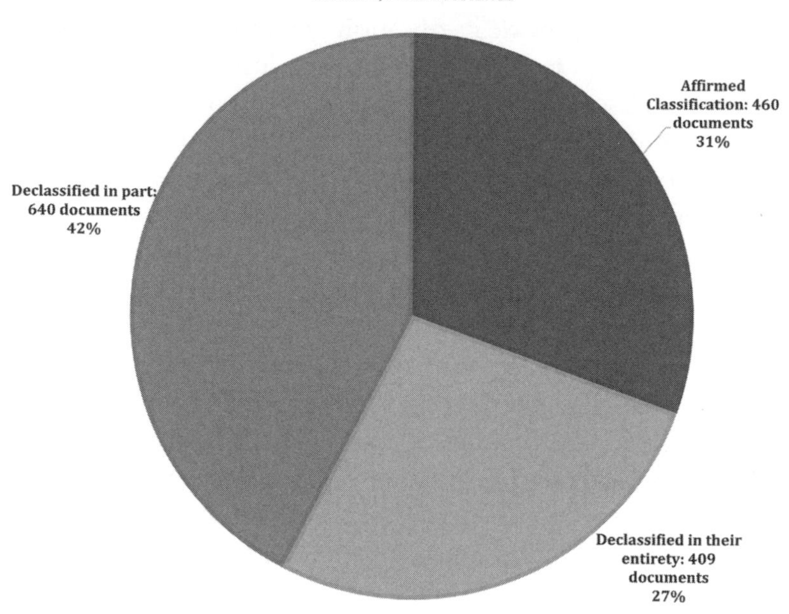

FIG. 20. ISCAP decisions on declassification, May 1996–September 2013. Source: Information Security Oversight Office, "2013 Report to the President."

unclassified based on declassification by ISCAP remain unavailable to the public. The CIA often cites the Central Intelligence Agency Act of 1949 (50 USC § 403g) to justify its authority to withhold many of its own unclassified documents.[43] The extensive redactions in figures 22 and 23 are strong illustrations of how declassification could be numerically counted as "successful" but where the spirit and intent of reducing secrecy are not achieved. One of the documents is four pages long, and yet only three sentences are left unredacted.

Increasing Secrecy

The trend toward increasing secrecy is also found in patent secrecy orders. As discussed in chapter 2, the number of secrecy orders is increasing over time. A 1993 government report by the DOD Acquisition Law Advisory Panel found that the number of secrecy orders was "excessive" and cautioned that increased secrecy has detrimental

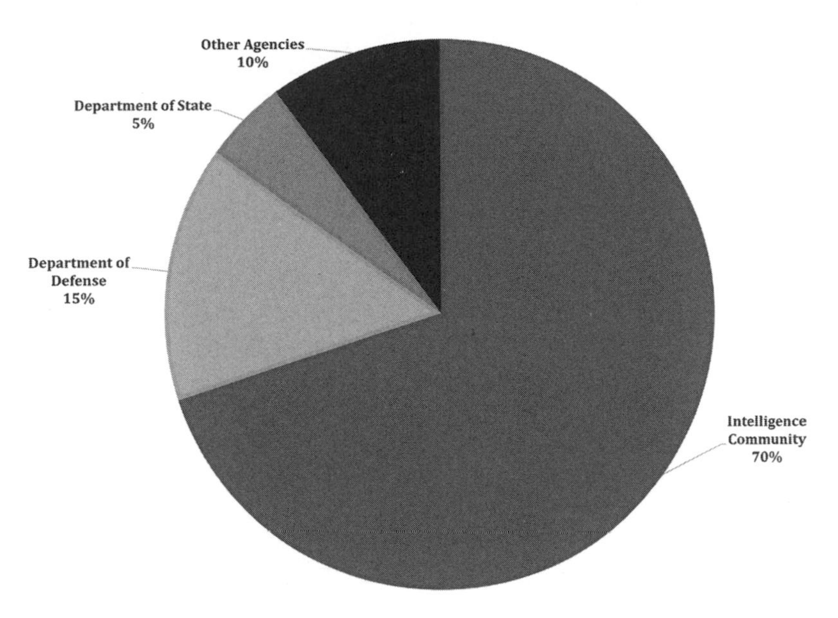

Fig. 21. Appeals to ISCAP. Source: Based on data from Interagency Security Classification Appeals Panel, "NARA and Declassification."

effects on innovation: "The process being used at the present time places many patents under secrecy order, thereby impeding the owner of the invention from using it in worldwide commerce."[44]

The ramifications of a patent secrecy order must be understood. Not only does such an order prevent the inventor from obtaining a patent, it also forbids the inventor from publishing or disclosing any material related to the invention. Further, export control laws include "information covered by an invention secrecy order" in the definition of technical data that are restricted from export.[45] A secrecy order classifies not only the patent, but also all of the data contained in it. Therefore care must be taken to determine the process for classifying patent applications.

Chapter 2 described a detailed process involving security classification guides to specify what pieces of information should be classified and at what level. However, patent secrecy orders do not follow such a process, and no such guidance exists. The government

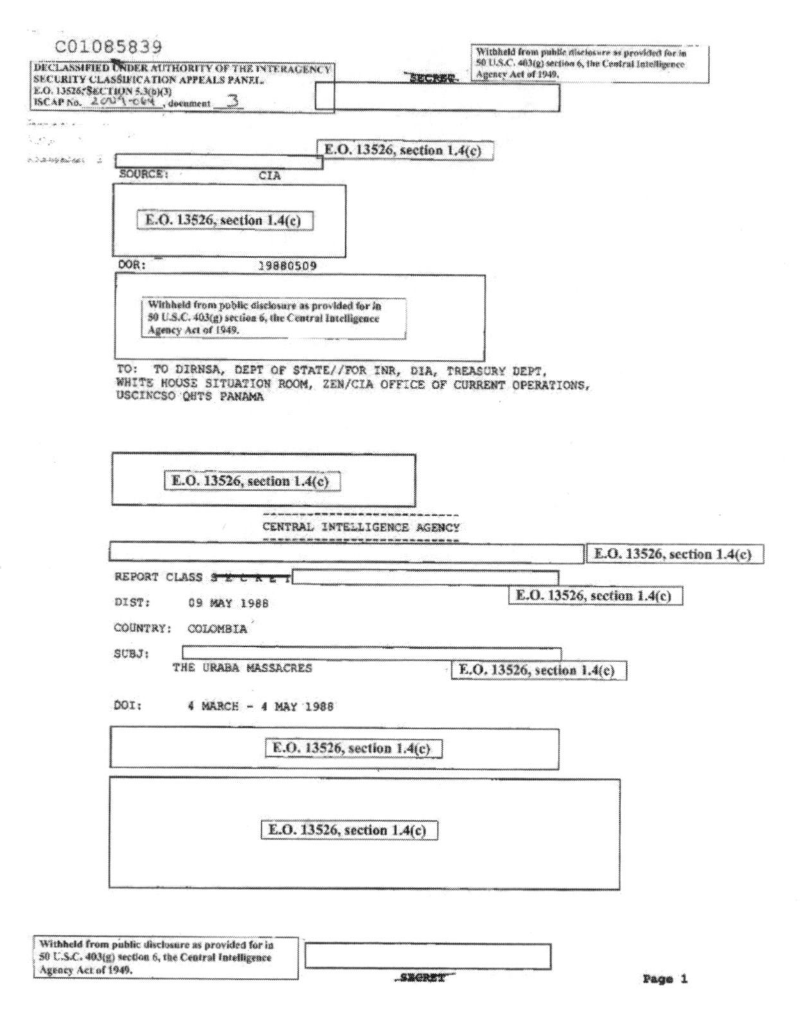

C01085839

DECLASSIFIED UNDER AUTHORITY OF THE INTERAGENCY SECURITY CLASSIFICATION APPEALS PANEL.
E.O. 13526, SECTION 5.3(b)(3)
ISCAP No. 2009-064, document 3

SECRET

Withheld from public disclosure as provided for in 50 U.S.C. 403(g) section 6, the Central Intelligence Agency Act of 1949.

E.O. 13526, section 1.4(c)

SOURCE: CIA

E.O. 13526, section 1.4(c)

DOR: 19880509

Withheld from public disclosure as provided for in 50 U.S.C. 403(g) section 6, the Central Intelligence Agency Act of 1949.

TO: TO DIRNSA, DEPT OF STATE//FOR INR, DIA, TREASURY DEPT,
WHITE HOUSE SITUATION ROOM, ZEN/CIA OFFICE OF CURRENT OPERATIONS,
USCINCSO QHTS PANAMA

E.O. 13526, section 1.4(c)

CENTRAL INTELLIGENCE AGENCY

E.O. 13526, section 1.4(c)

REPORT CLASS ~~S E C R E T~~

E.O. 13526, section 1.4(c)

DIST: 09 MAY 1988

COUNTRY: COLOMBIA

SUBJ:
THE URABA MASSACRES

E.O. 13526, section 1.4(c)

DOI: 4 MARCH - 4 MAY 1988

E.O. 13526, section 1.4(c)

E.O. 13526, section 1.4(c)

Withheld from public disclosure as provided for in 50 U.S.C. 403(g) section 6, the Central Intelligence Agency Act of 1949.

SECRET

Page 1

Fig. 22. Declassified document. Source: U.S. Central Intelligence Agency, "The Uraba Massacres" (ISCAP appeal number 2009-064-doc3).

has previously recognized this challenge. The DOD Acquisition Law Advisory Panel noted that "neither the PTO nor individual service branches and intelligence services have issued clear or consistent guidance concerning procedures for determining which technologies deserve scrutiny."[46] The panel looked at the process through which various agencies submit recommendations on secrecy orders

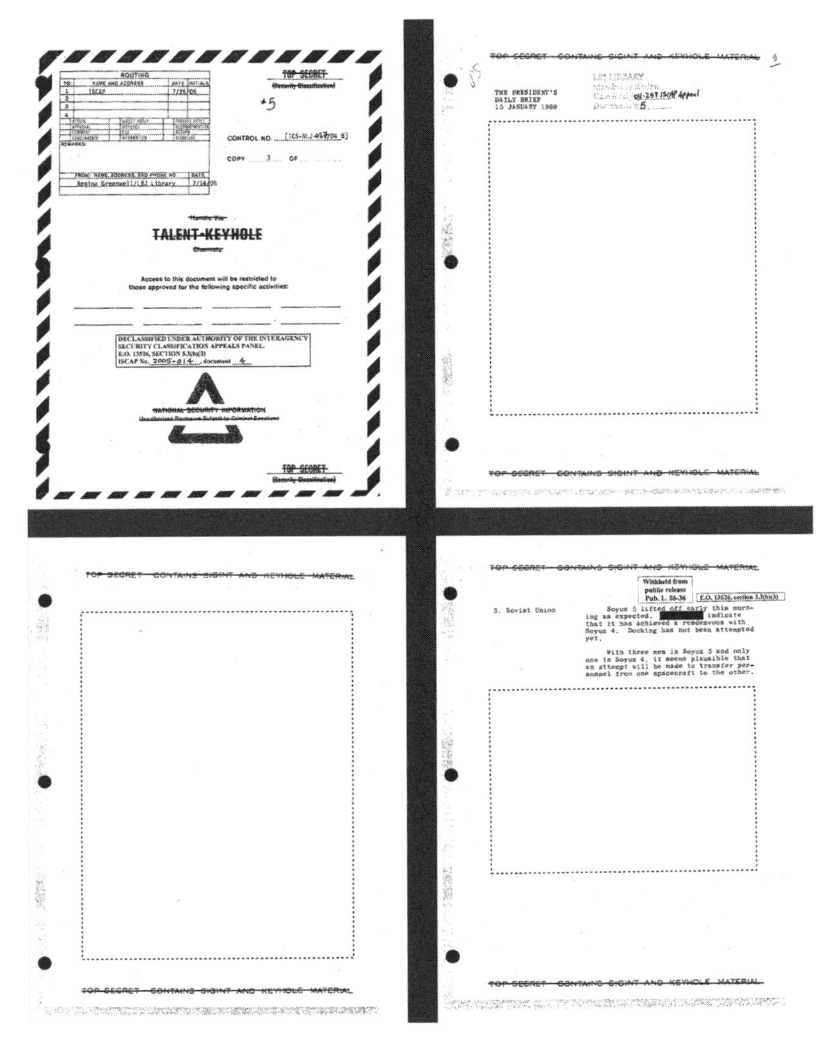

TOP SECRET
(Security Classification)

CONTROL NO. [TCS-NLJ-027/04 8]

COPY 3 OF

TO:	NAME AND ADDRESS	DATE	INITIALS
1	ISCAP	7/25/05	
2			
3			
4			

REMARKS:

FROM: NAME, ADDRESS, AND PHONE NO. — Regina Greenwell/LBJ Library — DATE 7/14/05

Handle Via

TALENT-KEYHOLE

Channels

Access to this document will be restricted to those approved for the following specific activities:

DECLASSIFIED UNDER AUTHORITY OF THE INTERAGENCY SECURITY CLASSIFICATION APPEALS PANEL.
E.O. 13526, SECTION 5.3(b)(3)
ISCAP No. 2005-014, document 4

NATIONAL SECURITY INFORMATION
Unauthorized Disclosure Subject to Criminal Sanctions

TOP SECRET
(Security Classification)

TOP SECRET CONTAINS SIGINT AND KEYHOLE MATERIAL

THE PRESIDENT'S
DAILY BRIEF
15 JANUARY 1969

TOP SECRET CONTAINS SIGINT AND KEYHOLE MATERIAL

TOP SECRET CONTAINS SIGINT AND KEYHOLE MATERIAL

TOP SECRET CONTAINS SIGINT AND KEYHOLE MATERIAL

TOP SECRET CONTAINS SIGINT AND KEYHOLE MATERIAL

Withheld from public release Pub. L. 86-36

E.O. 13526, section 3.3(b)(3)

5. Soviet Union

Soyuz 5 lifted off early this morning as expected. ██████ indicate that it has achieved a rendezvous with Soyuz 4. Docking has not been attempted yet.

With three men in Soyuz 5 and only one in Soyuz 4, it seems plausible that an attempt will be made to transfer personnel from one spacecraft to the other.

TOP SECRET CONTAINS SIGINT AND KEYHOLE MATERIAL

Fig. 23. Declassified document. Source: U.S. Central Intelligence Agency, "The President's Daily Brief" (ISCAP appeal number 2005-014-doc4).

to the PTO. It found that "agencies often rely on the Military Critical Technologies List (MCTL) to determine whether to recommend the imposition of a secrecy order to the PTO," even though "[this] list was never intended for such use."[47] The panel was concerned that using the MCTL "as a justification for the imposition of a secrecy order could cause severe constraints on the availability of critical technologies to U.S. defense industries by denying patent protection to U.S. technology innovators."[48]

Even where Congress has enacted statutes governing secrecy related to U.S. patents, the guidance is not clear and leaves open risks for misunderstanding, inconsistency, and potential conflicts of interest. The Invention Secrecy Act states that a secrecy order should be imposed when the disclosure of a patent "might, in the opinion of the head of the interested Government agency, be detrimental to the national security."[49] No clear guidance is given as to what would be "detrimental to the national security." Without such guidance, there is a risk that agency heads may apply their own, potentially varying, standards for what is "detrimental." In addition to the risk of inconsistency, there is a potential for conflicts of interest to arise when the same agency that may benefit from the use of or access to an invention is also the one in charge of ordering that it be kept secret. Since there are limited data published and such determinations are not made in a public manner, it is difficult to study these risks empirically or evaluate whether they are real or theoretical. But the uncertainty alone may be enough to deter some innovators from participating in secure government R&D.

Notwithstanding these concerns, the broader point remains the same: untangling secrecy and national security to achieve innovation can enhance national security in the long run. For example, if the determination of what is "detrimental to the national security" is not consistently understood or applied, it is difficult to argue that whatever national security interest might exist is truly being served by the law's protections. A more critical evaluation of what the national security interest is in a given situation, and whether such interest is best served by secrecy or openness, would likely better serve the national security interest. The DOD panel urged that the competing national security interests of innovation and secrecy need to be balanced: "The statute should operate in a manner that will promote the U.S. technological base while at the same time impede the flow of technologies to potential adversaries."[50]

In its 1993 report the panel recommended amending the Invention Secrecy Act and establishing an oversight committee to ensure that secrecy orders are applied more consistently and that both national security interests of secrecy and innovation be weighed. To date no amendment has been made.

Overcoming the Challenge

Untangling secrecy and national security to achieve innovation that enhances national security in the long run is a complex challenge. It requires enduring policies that can be carried out fully without their implementation being repeatedly examined and "corrected" through the interference of a wide array of government agencies. Yet such a procedure is simply not possible when policy is made through a series of executive orders that are reversed, one after the other, every time a new president enters the Oval Office.

The legislative and judicial branches can help break the logjam and provide guidance and oversight to ensure consistency. While the willingness of these branches to do so may be unclear when it comes to security classification specifically, they have shown a willingness to act in a related area of the law: disclosure of information. The passage and ongoing refinement of the Freedom of Information Act (FOIA),[51] enacted by Congress in 1966, is an instructive story.

The intent of FOIA was to allow the disclosure of information whenever possible. FOIA initially required disclosure unless it was specifically disallowed by another statute.[52] The Supreme Court later imposed restrictions on this broad disclosure, interpreting the law as having "a congressional intent to allow statutes which permitted the withholding of confidential information, and which were enacted prior to the FOIA, to remain unaffected by the disclosure mandate of the FOIA."[53] The court allowed the Federal Aviation Administration (FAA) to delegate "almost unlimited discretion to agency officials to withhold specific documents in the 'interest of the public.'"[54]

Congress responded by clarifying FOIA's intent. As the U.S. Department of Justice explains, "Fearing that [the Supreme Court's] interpretation could allow agencies to evade the FOIA's disclosure intent, Congress in effect overruled the Supreme Court's decision by amending Exemption 3 in 1976."[55]

What of the executive branch? The current pattern of executive orders is inconsistent and has serious limitations for addressing a complex issue that requires consistent, sustained, and directed action. The objectives behind executive orders on classification appear to be primarily concerned with transparency. They lack any explicit consideration of enhancing innovation in areas of importance to national security. Such specific consideration could help support other initiatives to reduce secrecy to the extent possible to enable that innovation. In the end all three branches of government need to work together if we are to realize the objective of less unnecessary secrecy in the interest of open innovation and national security.

SIX

Incentives for Innovation

A key challenge to ensuring participation in the innovation process for national security is providing adequate incentives to potential innovators. A large number of diverse innovators will help to enable, accelerate, and enhance technology innovation, which in turn will advance technological superiority and support national security objectives.

Incentives can be difficult because of the complex relationship between the U.S. government and innovators, be they individuals, teams, individuals within a company contracting with the government, or individuals within a company subcontracting to another company that contracts with the government. Analyzing the complex web of interrelationships reveals that the U.S. government may not be able to directly incentivize innovators and it does not exercise control over how intermediaries incentivize innovators.[1] The government must incentivize innovators to innovate and also ensure innovations reach the U.S. government to meet national security needs.

When innovators are part of small businesses that must work through established large prime defense contractors, the difficulties become greater—especially since the government does not have control over those relationships (as discussed in chapter 4). Implicitly requiring small businesses to work through these intermediaries has several side effects, including that prime defense contractors may put requirements on small businesses that can in effect preclude them from participating.

Even when innovators are part of companies already working directly with the U.S. government, incentives can be a challenge. Obviously the government can benefit from an innovation only if that innovation becomes known to the government. Such knowledge requires that the innovator employed at a company communicate the innovation's existence to the right people within the company and that the company then bring it to the government's attention. The latter can be impeded by government "overreach" (explained below) when it comes to securing rights to innovations that further national security. This "overreach" may further reduce innovators' incentives to continue innovating.

Since the government does not directly incentivize employees within a company, it must rely on the company to provide the appropriate incentives. Evidence shows that many contractors are unable or unwilling to provide adequate incentives to their employees; concerns include a lack of monetary support during invention disclosure processes, low monetary as well as non-monetary incentives for patents, and a lack of encouragement by management for employees to put forth inventions.

At most large defense contracting companies, employees must account for forty hours of work each week on specific, approved tasks. Submitting documentation as part of an invention disclosure process is typically not an approved task, so employees are left to disclose inventions on their "own time." On the other hand, when a company uses its internal funds to support an employee's research and development activities, like completing invention disclosure paperwork, it is called internal R&D (IR&D). Looking across departments of large prime defense contractors, few provide employees charge numbers to which they can allocate such time on their time sheets (which are used for billing the government as part of contracts). This burden on employees results in fewer inventions being properly filed, as evidenced by the high number of incomplete invention disclosures that did not result in either a patent filing or a trade secret. The primary reason cited for an incomplete invention disclosure was a lack of technical detail; this lack of detail accounted for nearly one-third of closed disclosures (see fig. 24). Nearly two-thirds of the invention disclosures that were not completed because of a lack of technical detail were cases in which an employee was required to develop the invention and complete the invention disclosure

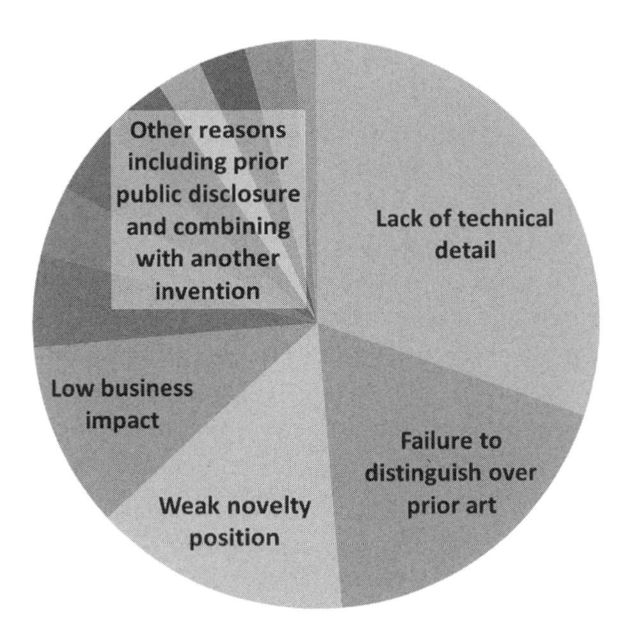

FIG. 24. Reasons for incomplete invention disclosure. Created by the author using information from defense contractor invention disclosure data.

paperwork on his or her own time without company support (fig. 25). Survey data confirm that employees consider the lack of time or funding support as the primary obstacles to submitting invention disclosures.

Even when there are monetary incentives for employees, they tend to be trivial compared to the revenue the company generates from an invention. Depending on the number of employees involved in a given successful invention disclosure (i.e., one selected for patent filing or as a trade secret), these monetary incentives typically range from a few hundred to a couple thousand dollars per inventor. Employees reported that these monetary incentives were not sufficient, in many cases, to motivate them to complete the process. Non-monetary incentives were considered unremarkable, such as a certificate of achievement or perhaps recognition at a company-sponsored event. Incentives also tend not to be tied to the significance of the national security need the innovation addresses or the financial incentives the company receives from the R&D contract or licensing.

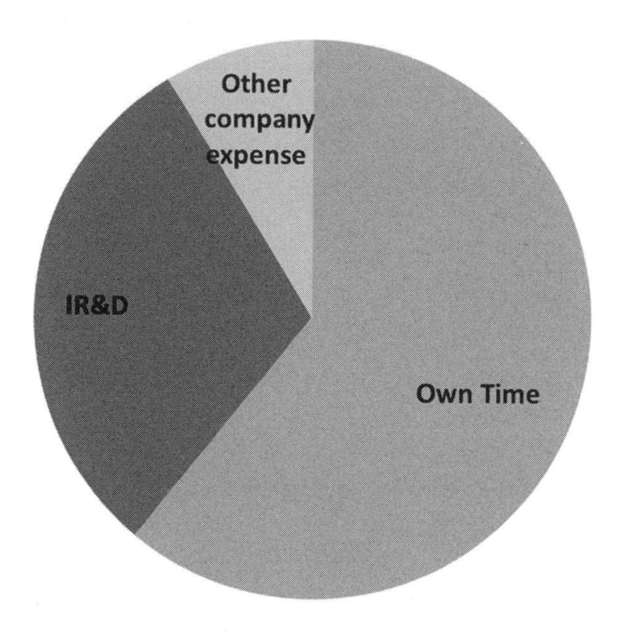

Fig. 25. Funding source for invention disclosures not completed due to lack of technical detail. Where government funding supported the employee, lack of technical detail was not cited as a reason for incomplete invention disclosure. Created by the author using information from defense contractor invention disclosure data.

Beyond the lack of financial support for employees, there can also be a lack of management support and encouragement of invention disclosure. Technology managers want their employees to be innovative, but they do not always support the process of identifying innovations to the company so that such innovations can in turn be brought to U.S. government customers. My research into invention disclosures also revealed inconsistent support in terms of defense contractors assigning mentors to support innovators through the invention disclosure process; such support is relevant because the process can be lengthy (several years) and cumbersome, and there can be a perception of favoritism within the selection committees. Several employees spoke of feeling excluded because the selection committee had "favorites." Overall fewer than half the employees surveyed said that they are encouraged to submit invention disclosures as part of

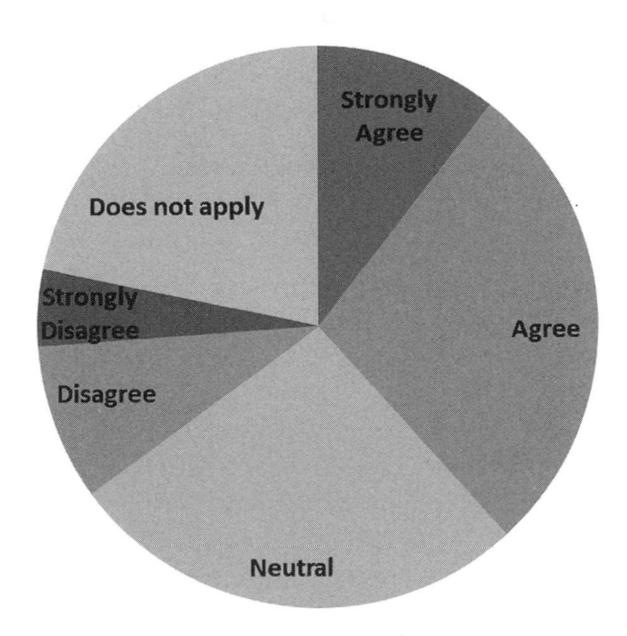

Fig. 26. Employee responses to the statement, "I am encouraged to submit invention disclosures as part of my job." Source: Created by the author using information from defense contractor employee survey data.

their job responsibilities (fig. 26). Many employees cited the lack of a charge number as a deterrent, as mentioned above, while others working on secure programs cited supervisors' concerns about leaking classified information, as well as a general lack of encouragement by the leadership of such programs regarding invention submission. Some employees called attention to the specific performance metrics used to calculate their bonuses and salary increases and the lack of a performance metric pertaining to intellectual property. Finally, some employees noted a preference by leadership to hold inventions as trade secrets rather than filing patents.

Such observations comport with my own experience. In 2011 I was working with a large defense contractor on a classified program, while also working on non-classified programs. My supervisors for the latter encouraged me to file invention disclosures and showed me how to do so through a company web portal accessible to all employees. With their support, I filed three.

The web portal specified, "Do not enter classified information," but it provided no direction to another process for filing a classified disclosure or indication of whom to ask. I had invented many things for the classified program, but my supervisors did not understand why I would want to file invention disclosures on those and discouraged me from doing so. In certain cases they were concerned that it would be a lot of effort to work with security to ensure that classified information was handled properly and could create unnecessary risk, and this effort would be a distraction from working on the program. In such situations I filed nothing.

Of course the government benefitted from the subset of the inventions that were incorporated in the technology we delivered. But by the company's choosing not to file a patent, the government never got information on *how* certain things from which it benefited actually *worked*—and thus was left unable to take those benefits further. In contrast, if the company had filed a patent, the company would have had to explain how the invention worked as required in the patent disclosure process. Inventions that were ultimately not implemented in the technology we delivered were not provided to the government. Thus potentially more effective solutions that were not implemented for other reasons, such as cost, were not disclosed. At some time in the future, when the forward march of technology made those alternative ideas cost effective, the government—and U.S. national security—would not have the benefit of that innovation.

Co-opting Inventions in the Name of National Security

The regulatory regime governing secure U.S. government R&D has afforded certain allowances to government agencies to co-opt or acquire innovations from innovators with the justification of national security. This allowance has significant effects on the ability to incentivize participation in the process of innovating in areas of importance to national security.

The U.S. government has recognized these effects. The DOD Acquisition Law Advisory Panel, commissioned by Congress in

1991 to review and streamline defense acquisition laws, noted that "commercial technology has outpaced DOD technology in a number of areas of vital importance to the development of weapon systems" and that technology owners have an "increasing reluctance to use their best commercial technology if there is a possibility that DOD will take the intellectual property rights in that technology."[2]

I explained the concept of classified patents, or patents protected by secrecy orders, in chapter 2. Given that secrecy orders prohibit inventors from publishing or disclosing any information related to an invention, classifying a patent essentially removes all or most commercial benefit a company could have derived from the innovative technology.[3] Congress provides a right to compensation for the inventor in such situations under 35 USC § 183: "a sum not exceeding 75 per centum of the sum which the head of the department or agency considers just compensation for the damage and/or use." But even assuming that all parties agree to the amount that is "just" compensation, awarding only 75 percent of that amount can easily be perceived as less than adequate and may lead to companies either not innovating in national security areas or selecting not to disclose certain innovations to the government to avoid the risk that the government will classify the technology.

I have experience with this issue. I was a co-inventor on four classified invention disclosures while I was an employee of a large defense contractor. During an Invention Review Committee meeting, we inventors were told that the disclosures met the criteria for patentability, except that the subject matter was classified. One committee member said, "Classified patents are useless because they are not enforceable until they are declassified. They are usually declassified only when the invention is useless now anyway." It was a catch-22. Thus there was a sense that there was no commercial value to a classified patent. Since we were competing with others in the industry, the committee decided to keep the inventions as trade secrets and not file them as classified patents. As inventors, we received about $1,000 per disclosure.

Authorization and Consent

The government allowances to co-opt innovations for national security, which reduce incentives for innovators to participate in the process, go beyond classified patents. They are wrapped up in what is known as "Authorization and Consent," a law that ensures ready U.S. government access to key national security technology inventions. The law provides statutory immunity to government contractors, absolving them of liability for infringing competitors' patents when done at the direction of the U.S. government (whether that direction is express or implied).[4] Authorization and Consent (A&C) is codified in the law as 28 USC § 1498, along with its corresponding FAR Clause 52.227–1.

The purpose of this World War I–era law was "to facilitate military and other governmental missions."[5] Indeed when Authorization and Consent was first conceived, U.S. Supreme Court chief justice William H. Taft recognized that the protection was needed to "stimulate contractors to furnish what was needed for the War, without fear of becoming liable themselves for infringements."[6]

It is important to understand the *century-old* environment in which the Authorization and Consent law was enacted. The United States had recently entered World War I, which had been raging for nearly half a decade. It was the bloodiest war anyone had ever seen. In the spring of 1918 the Germans had launched a major offensive on the Western Front, a stunning move that shook the confidence of the Allies in their ability to win the war. Then, in the middle of the war, a defense contractor with a patent on turbines (fig. 27) sued another defense contractor, seeking an injunction to keep its patented technology from being used.[7]

Photographs from the era (figs. 28 and 29) of specific ships cited in court proceedings that fought in World War I will reinforce in readers' minds just how much time has passed since Authorization and Consent was the "answer" to the government's problem and how different things are in the twenty-first century.[8] As the photographs show, the United States was waging war with ships powered by steam turbine engines.

The District Court denied the infringement claim, but then the

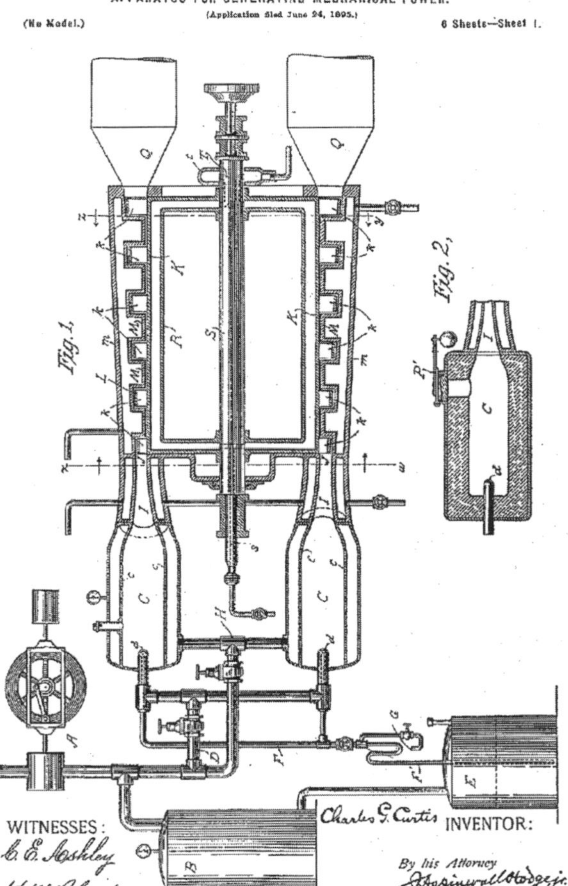

FIG. 27. Apparatus for generating mechanical power (Patent No. 635,919, patented October 31, 1899). The patent description by inventor Charles G. Curtis states the following: "A side sectional view of my improved combustion-engine and showing my new method of developing power."

FIG. 28. USS *Aylwin* (Torpedo Boat Destroyer No. 47), circa 1916–17. Source: Naval History and Heritage Command.

Circuit Court of Appeals reversed the lower court—thus affirming the infringement claim—but did not allow an injunction. The case made it all the way to the U.S. Supreme Court, which affirmed the ruling of the Court of Appeals regarding the infringement and also stated that a 1910 act of Congress written to protect patent owners did not preclude an injunction. The case was remanded back to the District Court to reconsider an injunction.

The established right of a patent owner to seek such an injunction had never before been tested at such a pivotal point during a war. If granted, the injunction would stop the supply of critical turbines to the United States as it sought to fight the enemy. Before that could happen, then secretary of the U. S. Navy, Franklin Delano Roosevelt, wrote an urgent letter to the Senate requesting a statutory amendment to prevent injunctions against defense contractors.[9] In unusually swift action, Congress enacted the requested amendment within weeks.

That amendment represents the beginning of Authorization and Consent. Enacted on July 1, 1918, in the heat of wartime (but not limited to wartime), the amendment states that the sole remedy against either the U.S. government or a defense contractor

FIG. 29. USS *Warrington* (Destroyer No. 30) under construction at the Cramp Shipyard, Philadelphia, April 1, 1910. Source: Naval History and Heritage Command.

is a suit against the U.S. government. It laid the groundwork for statutory immunity—in other words, that government contractors can intentionally infringe patents. And, it turns out, Authorization and Consent can have a devastating effect on the patent holders who have invested in R&D to develop the innovations at the heart of our nation's security.

The Authorization and Consent law is not limited to projects involving national security, and "the Government has exercised this right very widely—giving authorization and consent to use private patents on almost all Government contracts."[10] When Authorization and Consent is included in contracts, the contractor can "intentionally infringe patents in the context of government contract work without incurring any liability," while the patent owner is left to pursue a claim for "reasonable compensation" against the government in the U.S. Court of Federal Claims but cannot seek an injunction or obtain any remedy from the infringer.[11]

From a patent holder's perspective the sole remedy of "reasonable compensation" is a significant limitation on the remedies otherwise available for willful patent infringement in a traditional patent case. Under 35 USC § 271, the patent holder in a traditional patent case can seek an injunction preventing continued infringement and can seek damages in the form of a reasonable royalty, lost profits, and even treble damages and attorneys' fees for willful infringement. Under Authorization and Consent, though, patent owners may not be aware that their inventions are being used. If they do find out, they face high legal fees to obtain compensation (which is limited because they cannot recover damages or their legal fees). Therefore Authorization and Consent can be a disincentive to innovate in national security areas. The disincentive can be seen clearly in the story of Tenebraex, a small company that developed an anti-reflective coating for rifle scopes.

Tenebraex invested its own money; no U.S. government funding was used in the R&D. The company filed for and was granted patents on its invention. Then the U.S. Army found the coating to be essential for the military because it "eliminated the glare from rifle scopes and also prevented the scope lenses from reflecting light that would be visible to the enemy."[12] In fact Tenebraex's anti-reflective coating was so useful that the army solicited quotes for machine-gun sights with the same coating and even cited the exact specifications in Tenebraex's patent.

Tenebraex decided to submit a quote in response, partnering with Elcan Optical Technologies, a company experienced in delivering scopes to the army and a subsidiary of Raytheon, one of the Big Five defense contractors. But instead of Tenebraex's winning the contract, or even being selected as a subcontractor to Elcan, the army made an astounding decision: it awarded Elcan the contract, bypassing Tenebraex altogether. Elcan/Raytheon manufactured the machine-gun sights by openly infringing Tenebraex's patent.[13] "Our procurement," said the counsel of the Army Materiel Command, "is based on [getting] the best product for the best price, not whether someone has a piece of paper saying he has intellectual property rights."[14]

Tenebraex's CEO was devastated. Describing the impact of Authorization and Consent, he said, "An infringer will always win a competitive bid because the poor sap who invented the technology has to include the often substantial cost of his R&D in his bid price, while the infringer does not. And even if the patent holder after great expense and effort secures a royalty from the Federal Government, that royalty . . . does not pay for employees, ongoing operations or more R&D."[15] The competitive advantage of the infringer was also observed by the DOD Acquisition Law Advisory Panel, which recognized that "the infringer can offer a price which does not include recovery of the costs of making the invention."[16]

Tenebraex, like many other companies, had little choice but to spend hundreds of thousands of dollars in legal fees just to obtain a modest compensation from the government for use of its patented technology by a competitor. The result is that companies can be "discouraged [from] figuring out ways of keeping our soldiers safer."[17]

Authorization and Consent creates a tension between two goals: on the one hand, the encouraging of robust and open innovation—that is, the finding and development of the best, most innovative technologies to further the government's mission—and, on the other hand, national security, which requires the government to have access to critical patented technologies. This tension is not trivial to resolve. It is just as important to ensure participation by contractors who did not invent the critical technology but who may be the best suited to deliver it (as noted by Justice Taft) as it is to ensure the participation and continued innovation of the inventing company so that future critical technology is actually developed and brought forth to the government. That is, both national security and innovation interests favor enabling broader participation. The ideal situation, from both the innovation and national security perspectives, would be to incentivize participation from both types of participants: the innovators and those best suited to deliver the innovation. Figure 30 summarizes this.

The current Authorization and Consent system does not always achieve the national security goal. Instead it creates additional tensions with innovation by placing control over innovator participation with the government agency overseeing a contract. Federal regulation 48 CFR 27.201-1 requires that the Authorization and

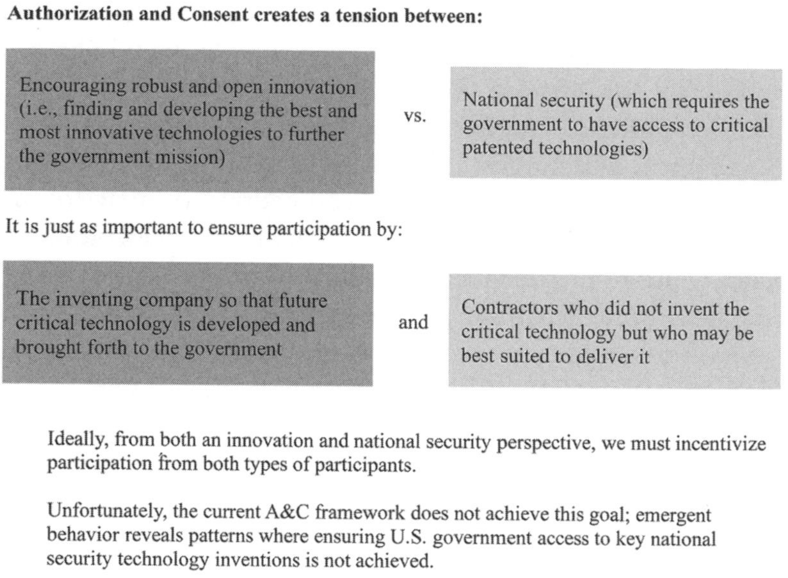

Authorization and Consent creates a tension between:

Encouraging robust and open innovation (i.e., finding and developing the best and most innovative technologies to further the government mission)	vs.	National security (which requires the government to have access to critical patented technologies)

It is just as important to ensure participation by:

The inventing company so that future critical technology is developed and brought forth to the government	and	Contractors who did not invent the critical technology but who may be best suited to deliver it

Ideally, from both an innovation and national security perspective, we must incentivize participation from both types of participants.

Unfortunately, the current A&C framework does not achieve this goal; emergent behavior reveals patterns where ensuring U.S. government access to key national security technology inventions is not achieved.

FIG. 30. Tension created by Authorization and Consent. Created by the author.

Consent clause be included in contracts, except when the complete performance and delivery are outside the United States or when simplified acquisition procedures are used, situations that represent only 2 percent of contracts.[18] There are several variants of the clause that offer varying degrees of protection to contractors. The most favorable variant to a potentially infringing contractor is Alternate 1: "The Government authorizes and consents to all use and manufacture of any invention described in and covered by a United States patent in the performance of this contract or any subcontract at any tier."[19] In addition to selecting which variant to include, contracting officers also have the discretion to include the accompanying patent indemnity clause, which "transfers liability to compensate the patent owner on to the contractor."[20] Further, contracting officers can also include a waiver of indemnity for specific patents the contracting officer identifies.[21]

Therefore, when an agency "wants patented technology, but does not want to pay the premium associated with contracting with the patent owner," the agency can include the Authorization and

Consent clause variant in the contract with the most protection for the infringing contractor and write contract specifications that make it necessary to infringe the patent to perform the contract.[22] Such an approach hurts the patent owner while encouraging participation by low bidders. If the agency does not want to invite full competition and instead wants to work directly with the patent owner, the agency can include a more restrictive variant of the Authorization and Consent clause and include the patent indemnity clause, thus deterring competition for the bid "without having to address public concerns about favoritism and economy."[23] Competition is deterred in such a situation because bidders are forced to internalize the cost of potential infringement through the indemnity provision that shifts liability back to the contractor.

It is not clear whether such manipulation over participation advances national security interests or how widespread this practice is, but it is certainly difficult to detect, measure, or deter under the current system. Approaches that limit such agency discretion and provide a more purposeful, directed framework would be helpful.[24] Such approaches would require addressing the degree of discretion government agencies have over which variant of the Authorization and Consent FAR clause is included in contracts, whether to include the patent indemnity clause, and whether to include a waiver of indemnity for specifically identified patents. Congress should provide guidance to agencies on how they must apply the Authorization and Consent clauses in contracts to level the playing field and increase participation by key players in areas of importance to national security.

One principle that emerges from evaluating this issue is that the tension is perhaps created more by the government's unwillingness to compensate for innovation adequately rather than from the tension between national security and innovation. That is, it is understandable that both the innovator and the contractor best suited to deliver the innovation should be incentivized to participate, and therefore immunity for the delivering contractor may be necessary to balance interests. Problems arise, however, because the Authorization and Consent rules seem to protect the contractor that delivers the innovation over and at the expense of the innovator. This does not mean that the innovator is not compensated (recall that the patent holder can seek reasonable compensation by filing suit in the U.S. Court of Federal Claims) but that the patent holder is not compensated sufficiently to ensure a continued incentive to participate. Thus the challenge becomes finding the answer to these questions: What is reasonable compensation? What type of compensation system is needed to ensure continued participation from innovators?

The feeling of property owners that they have not been justly compensated when their property has been taken has long been an issue; the taking of land through eminent domain may be the most familiar example. Case law shows that patent owners such as Tenebraex are not alone in feeling inadequately compensated for the "taking" of their intellectual property under the Authorization and Consent system.[25] The issue takes on additional significance when it results in reduced incentives for inventors to innovate in areas of importance to national security. For example, in *Robishaw Engineering v. United States*—a dispute related to patent license negotiations between the U.S. Army and a patent holder—the District Court focused on an entitlement to "reasonable compensation" rather than a right to be "made whole" (a type of remedy requiring compensation to fully account for a wrong).[26] While the inventing company got only "reasonable compensation," the competitor that was selected was "able to generate revenue, earn profits (which would never be disgorged), and build business relationships that would give it an economic advantage in future competition with [the inventing company]."[27] Similarly the Federal Circuit Court held in another case that a limited royalty of 10 percent was reasonable even though it was less than the "full compensation" that a traditional suit would have provided.[28] In another Federal Circuit case the court held that "a reasonable royalty is not equivalent to either the patent owner's lost profits or the cost savings realized by the government in selecting a different contractor than the patent owner."[29] In each situation the innovating company was limited to "reasonable compensation" and missed out on the additional economic upside of its inventions.

These cases show that "reasonable compensation" is simply not sufficient as an incentive for a company to remain in the business of innovating for the U.S. government. Unless the innovating company can reap benefits such as ongoing profits and business relationships, it will not have an incentive to continue to compete for government contracts. From a policy perspective, the goal should be not merely legally adequate compensation, but rather the provision of the compensation necessary to ensure a continued incentive to innovate. Such a policy would also be aligned with the economic reality that innovation is not free and that the cost of R&D must be internalized somewhere.

The current system does not force the government to internalize innovation costs into its pricing and instead offloads those costs onto the innovator. The consequence should not be a surprise: innovating companies lose their incentive to continue to innovate while their competitors enjoy a huge incentive to leverage those same innovations—which are not their own—without having to internalize the costs.

As of this writing, Congress has yet to amend the statute. As a result, even competing contractors have acknowledged the beneficial position in which it puts them. For instance, as Elcan/Raytheon patent counsel William Schubert wrote to Tenebraex, "Due to the Authorization and Consent clause of the Federal Acquisition Regulation 52.227–1, we have no need to examine either the coverage of the referenced patent or its validity."[30] That sentiment is not unique. Key personnel at defense contractors told me in interviews that the Authorization and Consent clause has affected their business models and R&D investment areas. As one general counsel of a Big Five defense contractor put it, "Authorization and Consent makes patents useless for defense contractors."

In summary, the Authorization and Consent clause may be discouraging participation by innovators in the innovation process for areas important to national security. Because the U.S. government is not forced to internalize the cost of innovation, there are consequences for long-term access to innovations that support national security. The Authorization and Consent system could be said to be compromising long-term national security success for short-term gain. Taking an invention to solve a near-term national security problem may be important, but doing so without providing corresponding compensation to the innovator may well result in a future reduction in the number of innovations available to the U.S. government.

The Authorization and Consent Workaround

The government's national security interests are best served by access to innovations and the ability to select the companies best suited to deliver those innovations to meet national security needs. Some companies, however, have taken measures into their own hands and find workarounds to let them hold on to their innovations. The most common method I came across is for innovating companies to designate innovations as trade secrets (a practice that sometimes requires creative accounting). This practice jeopardizes the government's national security interests because it prevents the government from having the innovation delivered by any other source. I call this practice the Authorization and Consent workaround. Some contractors work around the bind in which Authorization and Consent places them by keeping innovations in house—meaning they do not allow the core innovation or "secret sauce" to be disclosed through a patent or in some other form to the government. Training courses offered by prime defense contractors to their employees even include information about the trade secret option and its benefits over patenting given the Authorization and Consent regulatory regime. One such IP training course notes the following about Authorization and Consent:

> It means we have to actually sue the government in order to get paid for our invention. It is not something we like to do—sue the government. So this Authorization & Consent statute is a significant limitation to the power of patent enforcement in some of the markets where we serve. What this means is that [Company A] can openly copy a [Company B] patented technology in a fighter jet that they deliver to the U.S. government and we cannot use patent law to stop that infringement. On the other hand, should [Company A] employ this same technology in a jet delivered to a commercial customer, they would have to obtain a patent license from us or face the consequences of a patent suit.

Of course companies cannot withhold access to innovations if the government paid for the R&D that led to the innovation. But if the innovation resulted from either internal R&D funds or the efforts of an employee working independently (on his or her "own time"), the company is afforded stronger IP rights. The U.S. government generally receives only "limited" or "restricted" rights to intellectual property that defense contractors develop "at private expense," as negotiated in a U.S. government "IP Assertion Table" in all government contracts (these terms are common in U.S. government contracting).[31]

Four of the Big Five defense contractors receive greater than 80 percent of their sales revenue from the U.S. government, and they each spend less than 5 percent of sales revenue on internal R&D (the exception is Boeing, which has high commercial sales).[32] But some defense contractors are notorious for using "creative accounting" to show that most of their employees' inventions were developed at private expense—as figure 31 shows—even though those same employees spend the vast majority of their time working on government contracts.

The Authorization and Consent workaround illustrates how defense contractors can choose not to participate in the process of innovating for the U.S. government, a choice that does not serve the interests of national security. It is a direct result of flawed incentives. Efforts to overcome it would benefit from a focused analysis on the Authorization and Consent incentive system, including the development of metrics and collection of data. Such an investigation could reveal the broader and deeper effects of the workaround practice and could help identify solutions to correct it.

Government Handling of Proprietary Data

The struggle over incentivizing both types of participants—the inventing company and the company best suited to deliver the invention—has also played out in the handling of contractors' proprietary data. The 1947 Army Procurement Regulations favored the company best suited to deliver the technology; the government could take a company's proprietary data and share it with competitors. The 1955 Armed Services Procurement Regulation (ASPR)

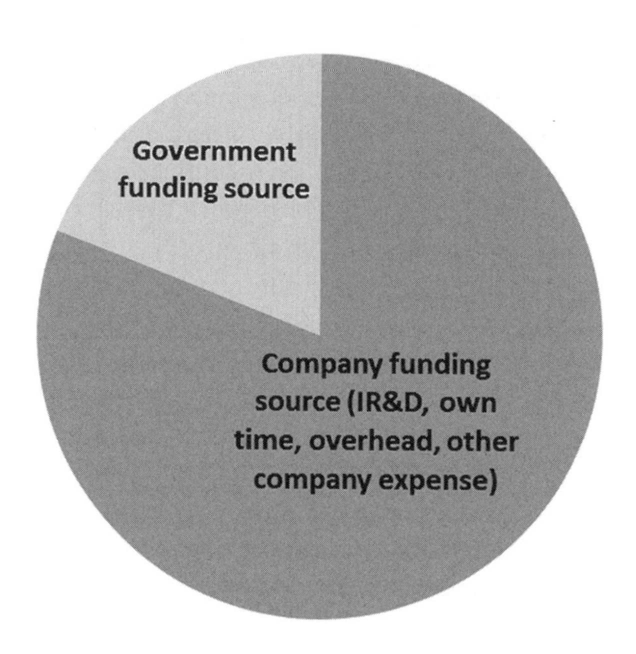

FIG. 31. Inventions categorized as being developed at the government's expense vs. private expense. Created by the author.

§ 9-202 favored the inventing company instead, stating, "It is the policy of DOD to encourage inventiveness and to provide incentive therefor by honoring the 'proprietary data' resulting from private developments and hence to limit demands for data to that which is essential for Government purposes." So an inventor would not have to say *how* the product it delivered to the government accomplished a task as long as the product accomplished the task successfully. Further favoring the inventing company, in 1964 the data rights policy changed to prevent the government from sharing proprietary data with competitors of the inventor. However, under public pressure regarding the high cost of military procurements, the pendulum swung back in the 1980s to favor the company best suited to deliver technology at a low cost. Although protecting contractors' proprietary data can bring innovators to the table, it can also lead to entrenched positions and can reduce future competition. Laws and regulations continue to evolve.

A 2006 court case highlights the tension between incentivizing the innovator versus the company best suited to deliver the innovation. The case concerns a small business, Night Vision Corp., which developed prototype night vision goggles under the SBIR program and successfully met the government's requirement by expanding the field-of-view of the googles without compromising image quality. In *Night Vision Corp. v. U.S.*, the small business complained that though it successfully completed SBIR Phases I and II, it was deprived of Phase III, which instead was awarded to its competitor, an established defense contractor.[33] The success of the small business in the Phase I and II contracts was undisputed. Night Vision Corp. sued for breach of contract, claiming that the success of Phases I and II should have resulted in the Phase III contract award. The court, though, held that the government agency was able to award Phase III to a competitor of the small business. Technical data owned by the small business was provided to the competitor. This case is particularly surprising given that Congress had an intention and policy goal for the SBIR program to be favorable to small businesses.[34]

Government Overreach in IP Policies

The U.S. government may have overreached in how it set IP policies in its early-stage implementations of open technology innovation strategies. The IP policy of the DARPA FANG-I challenge, introduced in chapter 3, provides an illustration.

The baseline IP rights policy for the FANG-I challenge was that the U.S. government would retain "unlimited rights" to all designs, even those not selected or awarded prize money. If participants wanted to retain stronger IP rights, they would first need to understand the policy and the FAR and DFARS well enough to understand that they needed to submit particular documents about the specific IP they wished to retain for negotiation in advance of entering the competition and generating the IP. To ensure the particular documents were submitted correctly and to negotiate those rights with the U.S. government, a participant would most likely need to retain an attorney. Further, the IP policy stated that trying to assert IP rights could affect the U.S. government's evaluation of the potential participant. It is not difficult to imagine that these regulations dissuaded at least some potential innovators from participating in the process. In fact a 2017 GAO report found that IP concerns were dissuading companies from working with the DOD.[35]

A direct comparison of the FANG-1 IP policy and that of a challenge on the open innovation platform Innocentive (discussed in chapter 1) amplifies this point. Here is an excerpt of the original FANG-1 policy language:

> Offerors responding to this BAA requesting a procurement contract to be issued under the FAR/DFARS shall identify all noncommercial technical data and noncommercial computer software that it plans to generate, develop, and/or deliver under any proposed award instrument in which the Government will acquire less than unlimited rights, and to assert specific restrictions on those deliverables. . . .
>
> In the event that offerors do not submit the list, the Government will assume that it automatically has "unlimited rights" to all noncommercial technical data and noncommercial computer software generated. . . .
>
> In accordance with DFARS 252.227-7013 Rights in Technical Data–Noncommercial Items, and DFARS 252.227-7014 Rights in Noncommercial Computer Software and Noncommercial Computer Software Documentation, the Government will automatically assume that any such GPR [Government Purpose Rights] restriction is limited to a period of five (5) years in accordance with the applicable DFARS clauses, at which time the Government will acquire "unlimited rights" unless the parties agree otherwise. Offerors are admonished that the Government will use the list during the evaluation process to evaluate the impact of any identified restrictions and may request additional information from the offeror, as may be necessary, to evaluate the offeror's assertions.[36]

In sharp contrast, Innocentive explains its IP policy in plain English—a procedure that ought to lead to increased participation. Perhaps more important, however, is that Innocentive does not afford the sponsor (what it calls the "seeker") IP rights to any of the ideas generated by participants (the "solvers") that were not selected/awarded. Here is an example of Innocentive's IP policy language in its description of a "theoretical IP transfer challenge":

> This Challenge is a Theoretical-IP Transfer Challenge, meaning that Solvers must relinquish all rights to the Intellectual Property (IP) *for which they are awarded*. By contrast, Theoretical-Licensing means that the Seeker is requesting non-exclusive rights to use the winning solution. For these forms of a Theoretical Challenge, Solvers that do not win retain the rights to their solution after the evaluation period is complete. *The Seeker retains no rights to any IP not awarded.*[37]

This seems like a fairer policy and one that may encourage Innocentive solvers to be less risk averse in their proposed approaches. For instance, consider a hypothetical situation in which a solver has an innovative, risky design idea for FANG-1, one that has a low probability of winning because of the lack of clarity on the metrics for evaluation (discussed in chapter 3). Nevertheless, the solver wants to try. Under DARPA's FANG-1 IP policy, the solver submitting such a design would give up some IP rights regardless of whether any prize money was won. With Innocentive, a solver could feel safe submitting ideas irrespective of their likelihood of winning without fear of losing some of the IP rights. In other words, the Innocentive participant has nothing to lose. The DARPA FANG-1 participant has everything to lose. Rational participants may drop out or fail to submit riskier ideas in the latter case—to the detriment of national security needs.

Although DARPA leaders sought to develop a favorable IP policy to encourage broad participation, they were constrained to operate within the confines of the FAR and DFARS.[38] At minimum the IP policy should be understandable and clearly communicated to participants. The new prize competition law, 15 USC § 3719, should be amended to include the provision to disclose the IP policy to participants "in a clear and reasonable manner," as is required in the new crowdsourcing and citizen science law, 15 USC § 3724.[39]

As was the case with the government handling of proprietary data discussed above, Congress can set the policy for managing IP rights. Section 10 USC § 2320, Rights in Technical Data, states that the secretary of defense shall prescribe regulations regarding technical data rights. However, the statute does not explicitly say that innovators have the burden of listing exhaustively or losing their rights; placing the burden on the innovators is *how* the regulation was written, specifically DFARS 252,227-7013(e). In the example of Night Vision Corp., failure to adequately mark the data rights on the goggles was how the government justified sharing the small business's information with a competitor. This regulation benefits the government by putting the onus on the innovator; however, it disincentivizes participation. Thus there is an opportunity for Congress or the agency regulators to modify this regulation to better incentivize participation.

Fixing the Flawed Incentives

What should the U.S. government do to confront the challenge of incentivizing participation in areas of importance to national security? As with the tension between secrecy and open innovation, the answer will not be found within the confines of a specific open innovation strategy implemented in a particular organization or even a particular U.S. government agency. A broader approach is required.

Over the years the U.S. Court of Appeals for the Federal Circuit has issued decisions related to the Authorization and Consent clause that have significantly limited a patent owner's protection against infringement by government contractors by reinforcing the idea that "reasonable compensation" is not the same as being made whole for infringement. Rulings from the Federal Circuit have also permitted the government to escape liability altogether in certain cases. For example, in *Zoltek Corp. v. United States*, the government awarded a contract to Lockheed Martin that infringed a manufacturing process patented by Zoltek.[40] Zoltek sought compensation under the Authorization and Consent framework. However, the Federal Circuit determined that Authorization and Consent did not apply, even if there was infringement, because some of the manufacturing had been outside the United States (in Japan), so no compensation was owed.[41]

On the other hand, the Court of Federal Claims has shown movement toward closing some loopholes that may have previously allowed the government to escape liability, once even going as far as finding "implied" Authorization and Consent because the government had previously negotiated with the patent owner and therefore was found to have been aware of its obligation to pay royalties.[42] "The Court of Federal Claims, since 2008, has reacted to the Federal Circuit's strict application of immunity by discovering creative approaches for protecting patent owners."[43] For example, in a 2009 case brought against the U.S. government by Boeing over a contract award given to its competitor, Lockheed Martin, the Court of Federal Claims took a broader view on reasonable compensation and awarded "the full compensation that [Boeing] could expect in a private infringement action."

Addressing the problem through the courts rather than through Congress or the agencies administering secure programs is not an effective, long-term, sustainable solution. The U.S. Court of Appeals for the Federal Circuit is not bound by decisions of the Court of Federal Claims, and each is limited in the avenues available to balance the needs of the patent owner and government. More important, their functions are to apply the law rather than to decide policy; it is up to Congress to weigh interests and develop a framework for incentives. While the Court of Federal Claims lacks the authority to provide non-monetary incentives, Congress could, for example, provide authority for the agencies to condition the award on the contractor's public acknowledgment of the inventor. That recognition may be valuable to certain innovators.

Congressional action is needed for success, given that Congress has the power to create necessary incentives to ensure participation from the right parties, and congressional action would be required to modify the existing Authorization and Consent law.

Several approaches that could strengthen the rights of the patent owner have been proposed. For example, awarding a mandatory 15 percent royalty has been proposed as an alternative approach to the long and expensive legal debates over reasonable royalty rates while ensuring a large enough incentive for innovators (as is done in a different U.S. law).[44] Although there are potential benefits to this approach (15 percent could be a large enough royalty to incentivize innovation), there are also drawbacks: so high a royalty could create difficulties for the government, and with some very large contracts involving multiple tasks it could lead to costly litigation regarding what portion of the total should be used as the base in calculating the royalty. Recent debates in commercial sector patent litigation have shown how hotly contested these royalty damages calculations can be (e.g., Apple v. Samsung smart phone wars over design patents).

Another approach that could strengthen the rights of the patent owner would be to enhance the due process or procedure required before the government can invoke Authorization and Consent. A variety of amendments to Authorization and Consent have been proposed but not implemented. One, for example, is that the government should notify the patent holder of infringement and the right to seek compensation. This proposal was not implemented because the government does not always know when it is infringing a patent. The question of infringement is itself not always clear. Also "because defense contracts typically have security classification and export control requirements, the government would often be able to claim that 'notification would be contrary to the public interest.'"[45] Another proposal was to limit Authorization and Consent to wartime. This was not implemented because the government can assert, at any time, that Authorization and Consent furthers national security.

To address the last concern noted above, and to narrow the application of Authorization and Consent more generally, Congress could limit it to programs critical to national security and then impose additional procedures that would require the government to justify the national security reason for using the patent (such as presenting it to an independent review board or court with appropriate secrecy assurances until a determination is made). The government could also be required to explore opportunities to bring an innovator into the fold to contribute to addressing the national security need or at least offering the inventor the opportunity to be heard on the issue.

Perhaps such a proposal would be analogous to the Foreign Intelligence Surveillance Act (FISA) courts, which oversee requests for surveillance justified on national security grounds. While views may differ as to whether FISA courts (or something similar) become "rubber stamp" overseers, at least such tribunals would offer some mechanism for protecting patent holders' interests with some "independent" oversight, which currently does not exist in the context of Authorization and Consent. Perhaps Congress would be willing to create a FISA court "equivalent." Now might be an opportune time for the recent public debate on reform of the FISA court system to extend further and potentially inform development of an analogous system for Authorization and Consent determinations, in turn helping to reduce uncertainty and encourage investment in innovation from both new and existing players to benefit national security. Unfortunately no fix appears on the horizon—even though Congress has known of the risks since at least the time of the 1993 DOD Acquisition Law Advisory Panel report.

Expand the Pie?

Modulating the level of IP rights demanded by the government is one way to further the goal of realizing technology innovation. Another is to go beyond IP rights to incentivize participation from innovators.

In and of themselves IP rights are not a goal. So why do potential participants want them in the first place? There are a variety of motivations, such as obtaining future business that derives from the invention, earning royalties from future contracts, and gaining a competitive advantage from being the one to bring about an innovation. What if the government could address these underlying motivations without compromising on the IP rights it feels it needs? It is not necessarily a zero-sum game.

Any recommendation on the IP rights for an open innovation effort cannot be made without understanding the entire incentive framework. First, the government must recognize the different types of groups that must be satisfied by the incentives. In general these may include private individuals and companies, as well as the employees within a company who are developing ideas and innovating. A system-level evaluation approach would be helpful, looking at the interdependence and interactive nature of elements of the system within and external to the government agency and the innovator organization. Such an approach could consider the full panoply of factors that influence an innovator or other demands on the innovator's time and energy—including, for example, that the innovator might be part of a company whose incentives are also relevant.

In that regard it is important to consider the employees within a company, not just the competing companies themselves. This consideration can be difficult because the government is generally not able to incentivize employees directly within a company and must rely on the company to provide appropriate incentives. It is also important to incentivize the company, such as with the promise of future business, or else the company may not allow an employee the flexibility to shift focus from a current project that is generating revenue to an innovation project proposal with no guarantee of success. Without such an incentive the company may be hesitant to provide the idea to the government.

There are also ways to reward participants even if they do not win. In the DARPA FANG-1 challenge a single team of three individuals won, but what about participants that contributed beneficial ideas and advanced the state of the art but did not win? Indeed the competition results show that some non-winning teams actually performed better on certain metrics than the winning team, but those metrics may not have been weighted as heavily.[46] Other prizes would be an incentive for participants to keep trying and innovating even when their chances of coming in first are low. Of course this option will require broader prize authority approvals that enable flexibility across open innovation strategy design variables.

It is also important to motivate participants to keep trying in future competitions. For example, a participant that does not win but performs better on certain metrics than the winner could receive a subcontract from the winner to contribute in those areas. Such an incentive structure, which could mimic the crowdsourcing incentive structures used by MIT and other competing teams in the DARPA Red Balloon Challenge (introduced in chapter 1), could also ensure that the final results for the government include the best solutions across metrics, not just those most highly weighted.

Participants might also be motivated to contribute in a beneficial way were the government to help them with security clearances, which is an onerous and expensive process, especially for smaller companies or new entrants to the defense contracting industry. It would be a recognition by the government that the participants show potential for contributing to national security projects in the future and a tangible demonstration of the government's interest in including them in future efforts. Such help is an example of an incentive that is not purely monetary and can have reputational and other benefits for the innovator.

Individual inventors and teams of inventors have specific needs (such as recognition, curiosity, patriotism), which are not necessarily monetary. Policymakers can benefit from in-depth analysis of these motivations and must recognize that incentivizing innovation does not need to be a distributive negotiation; instead we can expand the pie. Motivations go beyond IP rights as well. The debate about IP rights may be obscuring the underlying concern that participants must be incentivized to participate and that IP, while very powerful, is just one of the levers that can be adjusted to do that.

SEVEN

The Path to Long-Term Improvement

The United States exists in an unstable geopolitical climate. Each day the nation finds itself under threat of nuclear war, terrorism, cyber attack, and even social propaganda that affect how Americans live, work, travel, vote, and how we uphold our freedoms and democratic values. Our allies depend on a credible guarantee of American security forces. Perhaps more than ever before, those who lead U.S. national defense need to accelerate the pace of technology innovation, maintain a global technological lead, and keep ahead of our adversaries, who have shown they will stop at nothing to further their interests at the expense of ours.

Over the past century the United States has earned and maintained its status as the world's leader in technology innovation. This status has been earned through the undertaking of a variety of endeavors: from massive public works such as the Hoover Dam, to cutting-edge physics developing nuclear energy, to the Apollo program, which put people on the moon, and more recently to the tremendous innovation surrounding the internet and digital devices. For over a century the United States has led the way.

The United States' lead in global technology innovation and, more specifically, its lead in *secure* technology innovation is linked directly to the defense of the nation. Research in fields such as physics, mathematics, computer science, aeronautics and astronautics, machine learning, and other engineering subjects lays the foundation and basis for military transformation. Breakthroughs in technology—such as in radar, lasers, optics, nanomaterials, and microelectronics—continue to play a critical role in establishing and maintaining U.S. military superiority. A robust research portfolio surrounded by supportive U.S. government policies and regulations is a necessary part of a national security strategy that relies on the forward-looking knowledge that can anticipate threats and accelerate technology innovation breakthroughs to mitigate, foil, and outmaneuver our adversaries and keep the United States safe in a dangerous world. Therefore it is imperative that government organizations leverage the most advantageous technology innovation strategies for such programs (see fig. 1).

By successfully applying technology innovation strategies—and in particular open technology innovation strategies because of their demonstrated success—the U.S. government can enable, accelerate, and enhance the return on its investment in innovation. As the previous chapters have shown, however, the competing values that are endemic in secure U.S. government environments continue to create constraints and raise two fundamental challenges: the *secrecy challenge* and the *participation challenge*. The regulatory regimes governing these environments wield enormous influence on the adoption and execution of innovation. They simultaneously serve other values and interests that are in tension with innovation to varying degrees. Until these pervasive challenges are addressed, the U.S. government cannot optimize the application of open technology innovation strategies. Because these challenges are widespread across the defense system, they must be addressed through changes at the highest levels.

Surmounting the first challenge—secrecy—requires concerted action by Congress and the president. They are the only actors with the legal authority to address the secrecy challenge in a sustainable manner. Together the president and Congress must act to reform the regulatory regime governing secrecy and classification authority to ensure that government agencies tasked with bringing forth critical technology innovation are also empowered to do so in a way that thoughtfully balances the competing national security interests of innovation and secrecy.

It is a combination of legal and policy changes that will create the opportunities for sustainable improvement. The regulatory regimes must ensure adequate incentives for potential innovators to participate in the innovation process. That is the only way the government can ensure access to the most critical new technologies that support national security. Such access will require a thoughtfully constructed system of incentives, both monetary and non-monetary, combined with an appropriate internalization of the cost of innovating into the government acquisition process. Absent these reforms, short-term national security interests will continue to jeopardize long-term national security interests—even if unintentionally.

Such system-level change to the regulatory regimes governing secure U.S. government R&D is necessary to ensure sustainable improvement in the government's ability to transition R&D investment into technology innovation effectively. In fact establishing structures that support the optimization of technology innovation strategies into the regulatory regimes is itself a source of national competitiveness that supports national security.

It cannot be overstated: *the U.S. government will not be able to leverage open technology innovation strategies optimally to bring forth the most critical innovations to support national security without changing the regulatory environment*. The task is not easy, but it is important. Indeed it is *urgent*. Let's review how to get there.

Traditionally R&D investment is the primary driver of innovation, and it in turn supports technological superiority and national security. But in an increasingly multilateral world the United States cannot distinguish itself simply by outspending its competitors in R&D. Technology innovation strategies will become the differentiators that enable the United States to maintain technological superiority and national security by increasing the return on investment in innovation.

Open technology innovation strategies that have been applied successfully in the U.S. commercial sector could be potentially beneficial in secure U.S. government R&D environments. But the open innovation community has yet to make significant progress in adapting its strategies to these environments. The challenge is made greater by the fact that open innovation itself is not yet well defined as a strategy—as the innovation community acknowledges—and certainly not well defined in a secure government context. Adapting open innovation to the secure U.S. government environment will succeed only with the deepest possible understanding not only of the issues, but also of the very concept. For instance, as the analysis in this book shows, the term "open innovation" may be a misnomer because "open" strategies can be and are applied in "closed" environments as well—such as with classified innovator networks like NeedipedIA. Because private-sector companies provide innovation in traditional government R&D through contracting, these processes are already "open" in a sense. However, traditional U.S. government R&D contracting has been locked into a single open innovation approach for many years—one quite different from the open innovation strategies demonstrating breakthrough success in the commercial sector.

Overcoming Endemic Constraints

It is worth briefly revisiting four endemic constraints this book identified before summarizing the key recommendations for ensuring that the U.S. government has access to the innovation needed for national security going forward. Each of these constraints is a direct result of current regulatory regimes.

The first endemic constraint is the *pace and difficulty of maneuvering the bureaucracy*. The U.S. government R&D contracting bureaucracy has a negative impact on technology innovation. In serving competing values, the regulatory regime inadvertently constrains innovation through extremely complex and cumbersome regulation. Such constraints affect both large defense contractors and small businesses, albeit in different ways. Large defense contractors find themselves needing to make major investments in specific training and intellectual property attorneys just to navigate the FAR and DFARS codes. For small businesses, the regulatory regimes stand as significant barriers to entry. In one example it took more than three years from the time a determination was made by the agency that a start-up's technology met an urgent and critical national security need to the time that agency was able to establish the start-up on a government contract and pay for the critical technology—a time frame that was not unusual. In other words, the bureaucracy sets a pace that is at odds with the pace of start-up entrepreneurial activity, even when a need has been defined as urgent.

Further, the contracting mechanism is identical across a wide variety of government R&D contracts. That is, the bureaucracy required for contracts less than $1 million is similar, both in terms of the legal negotiations required and the actual contract template, to multi-billion contracts lasting decades. The vast set of clauses and legal provisions required for large contracts appears to be unnecessarily complicating the negotiation of contracts with small businesses, for which many of those provisions are irrelevant. Thus even if the barrier to entry is surmounted, the productivity and success of the small companies is called into question even before any actual contracted work begins.

The second endemic constraint is the *favoring of established players*. It is not uncommon in any contracting environment that established players that have previously obtained contracts have the advantage of ingrained relationships and ongoing work from those prior engagements. Such advantage subverts one of the key purposes of open innovation: to diversify participants in the innovation process. Policies that value experience in U.S. government contracting over the inclusion of diverse participants in the innovation process create lasting negative effects.

Beyond that, even when a smaller company might be allowed to participate, empirical evidence shows that U.S. government agencies favor and sometimes implicitly require that contracting companies have an agreement with a large prime contractor as a condition for award—thus further entrenching established players. In fact prime defense contractors are often intermediaries in the relationship between the U.S. government and small businesses. This role as intermediaries exists not by happenstance but rather as the result of policies. These policies can have a significantly detrimental effect on acquiring technology innovation for national security because they may, and often do, unintentionally exclude small businesses from the innovation process.

"SBIR shops" are another type of established player favored by the current regulatory regime. Although the SBIR program seeks to involve new participants, such as start-ups, a disproportional number of SBIR awards go to a handful of companies that have mastered the defense contracting game, detrimentally affecting the diversity of participants and the commercialization of SBIR-funded technology.

The perceived tension between national security and open innovation is actually a tension between *secrecy* and open innovation because of the assumption that national security and secrecy are inseparable. This is the third endemic constraint: the *secrecy challenge*. A rigid commitment to secrecy often precludes full and open competition, as well as information sharing—two factors important to the very innovation desired to serve national security interests. It is imperative that there be a genuine discussion about whether such secrecy truly serves the underlying national security interests. Without that discussion the commitment to secrecy at all costs will continue to limit participation and will keep secure U.S. government R&D projects siloed.

Finally, there is the *participation challenge*. This fourth endemic constraint is buttressed by the complexities introduced by having prime defense contractors as intermediaries in the relationship between the government and the inventor, and it contributes to the U.S. government's often being unable to incentivize innovators directly. Authorization and Consent, detailed in chapter 6, only makes this situation worse, as it adds to the lack of incentives the fact that the government is empowered to take patented technology without a license and use it for national security reasons, with limited liability. Such empowerment significantly reduces the incentives to participate in innovation important to national security. Further, Authorization and Consent contributes to the fact that certain defense contractors choose not to participate in the process and instead pursue workarounds—such as classifying innovations as trade secrets, thus protecting them from government "taking" and essentially defeating the government's own objective of having ready access to key national security technology.

Addressing the Challenges Sustainably

Two general principles emerge to overcome the secrecy challenge: do not overclassify and (if necessary) selectively declassify, and do not reduce competition unless necessary. These principles speak to the issue of disentangling national security and secrecy, and they are key to sustainable solutions. But U.S. government agencies that implement R&D programs are not sufficiently empowered with the authority to make these kinds of decisions; the president and Congress retain that authority but have not provided a solution. This needs to change.

With respect to overcoming the participation challenge, it is necessary to provide adequate incentives to ensure participation by potential innovators. Incentives are critical to having an innovation pipeline that supports national security. But it is not a matter of incentives, especially monetary ones, in a vacuum. The entire system needs to be reconfigured to mitigate the aspects that favor established players and erect barriers to entry for small businesses.

Specific recommendations include actions to be taken across the government in all three branches. The executive branch—including Cabinet departments, agencies, and even the Office of the President—needs to identify opportunities to simplify FAR and DFARS. The length and complexity of FAR and DFARS often fail to serve equal opportunity, non-discrimination, and other constitutional principles. They create barriers to participation. Further, there needs to be a new level of awareness that requirements in RFPs and contracts may help entrench the positions of established players, an outcome that runs counter to the diverse participation desired for open innovation. Moreover, it is time to recognize the importance of both monetary and non-monetary incentives. Congress should amend Authorization and Consent to include the provision of non-monetary incentives and a new kind of oversight—perhaps similar to the oversight established for the FISA process. Also related to the participation challenge, the new prize competition laws should be amended to require that the IP policy be disclosed to participants in a clear and reasonable manner, and it should be broadened to allow for the use of other open innovation strategies. In addition, it will be up to Congress to consider legislation that corrects the inconsistencies around secrecy classification policy. These are all necessary if changes are to be sustainable.

High Stakes

These recommendations will go a long way toward revamping how the U.S. government handles innovation in secure R&D environments and will thus ensure that the country maintains its ability to both secure its national interests and play its needed role in the world. The stakes are extraordinarily high, and the need for changes such as these cannot be overstated. The consequences of maintaining the status quo would be dire.

As noted at the beginning of this chapter, the United States exists within an unstable geopolitical scene. To navigate in this world, advance national interests, assure our allies, and play a role as a force for peace and stability, the U.S. government has historically used a combination of soft power and hard power—the proverbial "carrot and stick." For decades the United States has enjoyed an innovative edge in technology development. But as the rest of the world advances, that edge is diminishing. The simple reality is that losing our edge translates into a loss of hard power, which in turn translates into a loss of soft power, which in turn translates into an inability to influence matters that affect U.S. national security and interests. If American values are to be deployed to influence peace around the world, nuclear non-proliferation, climate change, humanitarian crises, and democratic rights—all tremendously difficult problems *now*—imagine how much more difficult they will be if the edge is lost.

That is the situation the U.S. government faces. There is a gap between the remarkable successes enjoyed in the high-tech commercial world *because of* how that world gets innovation done and how the government does it. And the gap is increasing. The situation creates an opportunity for other powers to fill the gap. We already see Russia deploying technology to expand its sphere of influence. China is showing a growing willingness to employ both soft and hard power globally, and the Chinese are rapidly building their innovation capabilities. Other countries will act in their own interests, not those of the American people. And if they become leaders in technological innovation, they will attract other countries through alliances that offer benefits, including protection and strength.

Failure to solve the innovation challenge the U.S. government faces will change who sets the world's agenda. The topics the United States wants at the top of the list for the United Nations, the G8, and other important international bodies will be replaced by those from whichever countries fill the leadership gap.

What happens when securing U.S. interests requires a specific technology we do not have and we cannot create quickly enough? We become reliant on others who are able to innovate and compete. The stakes go well beyond security and technological leadership and the power of military confrontation. If other countries fill the void, it could make for a world where American interests, culture, and values take a back seat. Today other countries rely on the United States to provide the technology to solve problems, such as determining whether a chemical or biological weapon has been deployed, finding a ship lost at sea, disarming pirates who have taken hostages, and so on. Many countries rely specifically on the U.S. military for defense and protection. When the need arises, they call the United States; in the future, they may not. If they do not, it may be because they can no longer rely on U.S. government innovation to develop the most advanced technology—and it may then be too late.

The case for change is compelling. The United States has taken dramatic steps to spur innovation before. When our national security, values, and way of life are threatened, we rise to the challenge—as the Apollo story illustrates. We put a man on the Moon. If we can recognize the crossroads we face, we will rise to the challenge again.

NOTES

Introduction

1. Takahashi, "Osama bin Laden's Death Reveals the Value of State-of-the-Art Technology."

2. Drew, "Attack on Bin Laden Used Stealthy Helicopter That Had Been a Secret."

3. Atherton, "New Details Emerge."

4. White House, "Maintaining Military Advantage."

5. White House, "Maintaining Military Advantage" (emphasis added).

6. "International Science and Technology Strategy for the United States Department of Defense," April 2005 (emphasis added). https://www.hsdl.org/?view&did=698413.

7. Silberzahn, "Conference on Strategic Surprises at the CIA."

8. Clinton, "National Security Science."

9. Clinton, "National Security Science."

10. White House, "Maintaining Military Advantage."

11. Task Force on American Innovation, "American Exceptionalism, American Decline?" (emphasis added).

12. Task Force on American Innovation, "American Exceptionalism, American Decline?" (emphasis added).

13. National Research Council, *Rising to the Challenge*.

14. INCOSE, "Systems Engineer: Guru or Scientist?"

15. Carter and White, *Keeping the Edge*.

16. Quoted in Doyle, "Outside the Box."

17. National Research Council, *Rising to the Challenge*.

1. The Emergence of Open Innovation

1. Chesbrough, *Open Innovation*, xxiv.

2. X Prize Foundation, "Story Ideas."

3. Ansari X Prize, "Mojave Aerospace Ventures Wins the Competition."

4. Quirky, "Everything You Need to Know About Quirky."

5. "How Quirky Turns Ideas into Inventions."

6. Spradlin, "Innocentive."

7. Birkinshaw and Goddard, "Combine Harvesting," 15–18.

8. Spradlin, "Innocentive."

9. Heredero, *Open Innovation in Firms and Public Administrations*, 177.

10. Lee, Hwang, and Choi, "Open Innovation." Also see Conrad et al., who note that "only a fraction of innovation challenges have been systematically evaluated" (*A Framework for Evaluating Innovation Challenges*, 1).

11. Two examples include Peer-to-Patent, a way to enlist the public to help find and explain prior art during patent examinations, and Katrina PeopleFinder, a service to find information on people affected by Hurricane Katrina. See Lee, Hwang, and Choi, "Open Innovation." In addition, Cleland, Galbraith, Quinn, and Humphreys note that "specific challenges of implementing Open Innovation in the public sector have not been adequately addressed" ("Platform Strategies for Open Government Innovation").

12. Chesbrough, "Open Innovation and the Design of Innovation Work."

13. Chesbrough, "Open Innovation and the Design of Innovation Work."

14. DARPA, "DARPA Today."

15. Hauser, Tellis, and Griffin, "Research on Innovation."

16. U.S. Securities and Exchange Commission, "Updated Investor Bulletin."

17. See von Hippel's work on user-centered innovation, such as *Democratizing Innovation*.

18. Ling, "User Centered Innovation Is Dead."

19. For further reading, consider concepts such "IP Modularity," discussed by Baldwin and Henkel in "Modularity and Intellectual Property Protection."

20. Currently the government reports only on its open innovation activities that follow the "Challenge" approach discussed above. See Office of Science and Technology Policy, "Implementation of Federal Prize Authority: Fiscal Year 2016 Progress Report."

21. The DARPA Shredder Challenge called upon participants to reconstruct shredded documents. Some have speculated that the true purpose of this challenge was to demonstrate a vulnerability in secure government shredding mechanisms rather than to develop a visual computer algorithm for reconstructing shredded documents because the government did not specify how to reconstruct the documents. Among the participants was a team from UCSD that attempted to use crowdsourcing to solve the puzzle. However, an individual sabotaged the team's efforts by logging in as one of the crowd and destroying all of the progress that was made in reconstructing the documents. Follow-up studies have been performed regarding crowdsourcing in an adversarial environment. Serbu, "DARPA Challenge."

22. Info Security, "DARPA Says Goodbye to Hacker Friendly Cyber Fast Track Program."

23. Gourley, "DARPA's Cyber Fast Track Adds Agility to Research Funding."

24. T. Webb, Guo, Lewis, and Egel, *Venture Capital and Strategic Investment*.

25. Gassman, Enkel, and Chesbrough, "The Future of Open Innovation."

2. Secret U.S. Government R&D

1. Sargent, "Federal Research and Development Funding: FY2013."

2. Fossum et al., "Discovery and Innovation."

3. Peck and Scherer, "The Weapons Acquisition Process," and Driessnack, "Unique Transaction Costs in Defense Market(s)."

4. The DOD publishes its R&D priorities in the *Quadrennial Defense Review* (*QDR*), a legislatively mandated review of DOD strategies and priorities. The 2014 *QDR* recognizes this complexity and its impact on innovation: "Successful innovation, particularly for an organization as large and complex as the U.S. military, is difficult. It will require strong, courageous leadership within the military, as well as close collaboration with our elected leaders" (U.S. Department of Defense, "Quadrennial Defense Review 2014").

5. Sargent, "Federal Research and Development Funding: FY2013."

6. Sargent, "Federal Research and Development Funding: FY2018," 5.

7. Proposal Exponent, "Federal R&D Funding."

8. Information with which to analyze this complex landscape is largely dispersed and fragmented within various divisions and departments of U.S. government IT infrastructures. Figure 9 is based on both publicly available and non-publicly-available data.

9. Small Business Research and Development Enhancement Act of 1992; U.S. Small Business Administration, "Legal Business Size Standards."

10. Small Business Research and Development Enhancement Act of 1992; U.S. Small Business Administration, "Legal Business Size Standards."

11. Priest and Arkin, "Top Secret America."

12. Priest and Arkin, "Top Secret America."

13. Department of Homeland Security, Office of Inspector General, "Reducing Overclassification of DHS' National Security Information."

14. Sensors Directorate, Air Force Research Laboratory, *Electronic Warfare Technology: Security Classification Guide.*

15. Sensors Directorate, Air Force Research Laboratory, *Electronic Warfare Technology.*

16. 50 USC–Trading with the Enemy, Act of October 6, 1917 § 10, CH. 106, 40 STAT. 411.

17. The Invention Secrecy Act, 35 USC § 181-88, and the related agency regulation 37 CFR, 5.1-5.5.

18. These restrictions are specified in 10 USC § 130, "Authority to Withhold from Public Disclosure Certain Technical Data."

19. Personnel and Document Security Division, "Protecting Classified Information."

20. 18 USC § 793.

21. "Eligibility Guidelines for Gaining Security Clearance."

22. U.S. Department of State, "All About Security Clearances."

23. The U.S. government agency would complete DD Form 254, which is central to all security clearance discussions and processes. The form identifies the government's requirements for security associated with a contract and specifies whether a facility clearance is required.

3. Success and Failure in Innovation

1. NASA Technology Transfer Program, "NASA Spinoff."
2. Scott, "Sputnik—50 Years Later."
3. Zacharias, "When Bomb Shelters Were All the Rage."
4. Cropley, *Creativity in Engineering*.
5. Dugan and Gabriel, "'Special Forces Innovation.'"
6. Kennedy, "John F. Kennedy Moon Speech."
7. Quoted in Brinkley, "50 Years Ago."
8. Quoted in Brinkley, "50 Years Ago," and "NASA Budgets."
9. Kennedy, "John F. Kennedy Moon Speech."
10. Johnson, "The New Space Race."
11. NASA, "Benefits from Apollo."
12. Lexington Institute, "Dragons of Change."
13. Axe, "Did the Marines' 40-Year-Old 'Amphibious Tractor' Just Strike Again?"
14. Government Accountability Office, *Amphibious Combat Vehicle*.
15. Sack, "Amos Rejects Recent Critique of Amphibious Combat Vehicle."
16. DARPA, "FAQS—Open Manufacturing," and Doyle, "Outside the Box."
17. Boyle, "How the First Crowdsourced Military Vehicle Can Remake the Future of Defense Manufacturing."
18. Suh and de Weck, "Modeling Prize-Based Open Design Challenges."
19. Fold It: Solve Puzzles for Science.
20. Cooper, *Predicting Protein Structures with a Multiplayer Online Game*.
21. 22 CFR § 120.3.
22. DARPA, "FAQS—Open Manufacturing."
23. DARPA, "Component, Context, and Manufacturing Model Library 2," 21.
24. 15 USC § 3719. This analysis does not presume that such restrictions are necessary or beneficial to national security but simply acknowledges that the restrictions limit the potential pool of innovators.
25. 15 USC § 3724.
26. DARPA, "Component, Context, and Manufacturing Model Library 2"; Innocentive, "Harvesting the Energy in Buildings."
27. Suh and de Weck, "Modeling Prize-based Open Design Challenges."
28. Eremenko and Wiedenman, Adaptive Vehicle Make (AVM)."
29. Elwell, "DARPA Offers $1 Million Prize for New Amphibious Armoured Vehicle Designs."
30. The prize authority concept has evolved from the Stevenson-Wydler Technology Innovation Act of 1980 (15 USC § 3719), the America Competes Act of 2007 (20 USC § 9801), the Prize Authority (NASA) (51 USC § 20144), and the Amer-

ica Competes Reauthorization Act of 2010 to two 2017 laws supporting government open innovation: 15 USC § 3719—Prize Competitions and 15 USC § 3724—Crowdsourcing and Citizen Science.

This recent 2017 legislation is a stride forward in government open innovation. However, the legislation echoes the narrow view the government has had of open innovation strategies; it follows commercial industry and is limited to two strategies: prize competitions and crowdsourcing. The strategies have further restrictions: the crowdsourcing legislation prohibits the financial compensation of participants. The prize competition legislation also restricts participants to "U.S. persons or companies."

Both new laws require agencies to prepare reports, and the contents are specified, such as identification of opportunities for future prize competitions. A broader view, such as the identification of opportunities for open innovation strategies, would encourage agencies to identify the appropriate strategy for each situation. I would also recommend that reports include challenges and issues so lessons learned can be shared.

31. DARPA, "After Successful Design Challenge Competition and Testing, DARPA Begins Early Transition of Adaptive Vehicle Make Technologies."

32. Freedberg, "Amos Says Marines to Drop High Speed ACV"; Sack, "Amos Rejects Recent Critique of Amphibious Combat Vehicle."

33. Snow, "The AAV Is Not Dead Yet." Although ultimately FANG did not continue, its parent program, Advanced Vehicle Make (AVM), was transitioned successfully into the national Manufacturing USA initiative, where its outputs have been leveraged to enhance the competitiveness of U.S. manufacturing in nonmilitary applications. Specifically AVM outputs have helped to digitize the supply chain with innovations in product design and systems engineering. AVM was transitioned into the Digital Manufacturing Design and Innovation Institute (DMDII), a cutting-edge, collaborative research institute that brings together industry, government, and academia to enhance the competitiveness of U.S. firms by accelerating innovation in manufacturing. DMDII is part of the national Manufacturing USA initiative, established by the Revitalize American Manufacturing and Innovation Act of 2014, 15 USC § 278s. DARPA, "After Successful Design Challenge Competition and Testing"; Manufacturing USA, "2016 Annual Report."

4. Consequences and Incentives

1. Author interview with John Akula in 2014. Akula is senior lecturer at the MIT Sloan School of Management and has primary responsibility for the school's business law curriculum.

2. Dwyer et al., "The Global Impact of ITAR."

3. Satellite Industry Association, "2017 SIA State of Satellite Industry Report."

4. Government Accountability Office, "Defense Contracting: Actions Needed to Increase Competition."

5. Calandrelli, "An Evaluation of Short Innovation Contest Implementation."

6. Obama, "Executive Order 13526: Classified National Security Information."

7. 42 USC § 1861.

8. DARPA, "DARPA Today."

9. Obama, "Executive Order 13526."

10. 50 USC § 3002.

11. National Science Foundation, "Small Business Innovation Research Program."

12. Kendall, "Public Access to the Results of Department of Defense–Funded Research."

13. Drach-Zahavy, *Understanding Team Innovation*.

14. National Science Foundation, "Small Business Innovation Research Program."

15. Gompers, *The Venture Capital Cycle*.

16. Gompers, *The Venture Capital Cycle*.

17. Calandrelli, "An Evaluation of Short Innovation Contest Implementation."

18. DARPA, "FAQS—Open Manufacturing."

19. Trimble, "First Foreign Firm Cracks U.S. Defense Industry's 'Big 5.'"

20. Excerpt from a "Big Five" training course on government IP regulations (emphasis added).

21. Anderson, "What Is the Difference in the DCAA and the DCMA?"

22. Brown, "Improving IP Protection for Small Businesses."

23. In 1990 BAE was known as FMC and later as United Defense (UDI), a "large firm [that made] combat vehicles and missile launchers for the U.S. military." When UDI was acquired by BAE in 2005, BAE became the "Pentagon's sixth-largest contractor" and "the second-largest maker of armored vehicles behind Falls Church, Va.–based General Dynamics" (Defense Industry Daily, "BAE Closes United Defense LP Acquisition").

24. Dalton, "A Case Study of the Advanced Amphibious Assault Vehicle (AAAV) Program."

25. "General Dynamics Land Systems, under their subsidiary General Dynamics Amphibious Systems, Woodbridge, Va., is being awarded a $712,026,417 cost-reimbursable contract. . . . This contract was not competitively procured" (U.S. Department of Defense, "Contracts").

26. Feickert, "Marine Corps Amphibious Combat Vehicle (ACV) and Marine Personnel Carrier (MPC)."

27. Freedberg: "Marines 2014" and "Amos Says Marines to Drop High Speed ACV"; General Dynamics, "General Dynamics to Continue Amphibious Combat Vehicle Testing."

28. Bacon, "Inside the Amphibious Vehicles That Won the Marines' $225m Contracts."

29. Judson, "BAE Wins Marin Corps Contract to Build New Amphibious Combat Vehicle."

30. Defense Acquisition University, "Omnibus Contracts."

31. Aitoro, "Small Businesses Often Lose in GSA Schedule Contracts."

32. DARPA Information Innovation Office (I2O), "Cyber Fast Track."

33. SBIR feedback letter, 2010. It should be noted that the small business had included a letter of endorsement from a government contractor, but the contractor was not large enough to count as a "major military entity, certified government prime."

34. A 2009 NDIA Small Business Division subcommittee survey showed that a vast majority of small businesses felt their IP was put at risk through the prime contract process. The subcommittee also found subcontractors being eliminated in the bid process if they "[didn't] agree to prime conditions on IP rights even if statutory rights should be non-negotiable" (Brown, "Improving IP Protection for Small Businesses").

35. Defense agencies "turned to the LSI concept in large part because they have determined that they lack the in-house, technical, and project-management expertise needed to execute large, complex acquisition programs" (Grasso, "Defense Acquisition"). However, the LSI concept faced criticism in many areas. For one, certain U.S. government agencies were becoming reliant on a single entity and single contract for the majority of their capabilities (e.g., U.S. Army on Future Combat Systems program, U.S. Coast Guard on Deepwater program). Another concern was a lack of transparency on selecting subcontractors: "In an LSI arrangement, the federal government has a contractual relationship with the LSI prime contractor, not with any subcontractors that report to the prime contractor" (Grasso, "Defense Acquisition"). Furthermore, programs such as Future Combat Systems experienced significant cost and schedule overruns and were eventually canceled (Srivastava, Arias, and Piper, "Future Combat Systems Case Study").

36. Starting in 2006, legislation was enacted to move the U.S. government away from the LSI concept: 10 USC § 2410p, "Contracts: Limitations on lead system integrators"; Loudin, "Lead Systems Integrators"; Young, "Lead Systems Integrator Role for Government"; U.S. Department of Defense, Office of Small Business Programs, "Guidebook for Facilitating Small Business Team Arrangements."

37. Zuberi, "SBIR/STTR Grants Are Great."

38. Zuberi, "SBIR/STTR Grants Are Great."

39. Small Business Association, "About SBIR."

40. DOD, "SBIR/STTR Program Desk Reference" (emphasis in original).

41. Data are available at SBIR.gov.

42. 78 FR 48537.

5. Secrecy versus Open Innovation

1. Obama, "Executive Order 13526."

2. Obama, "Executive Order 13526."

3. 42 USC § 1861. 81st Congress. (1950). Public Law 507—National Science Foundation Act of 1950.

4. Dahlander and Gann, "How Open Is Innovation?"

5. Terweisch and Xu, "Innovation Contests."

6. Frey and Haag,, "Whom Should Firms Attract to Open Innovation Platforms?"

7. DARPA sets time limits for technology development projects with the justification that innovators may be more likely to participate if they do not have to make long-term commitments. Shortly after they both left DARPA for Google, the former DARPA director and deputy director wrote in 2013 of this approach: "One of the most effective ways to attract talented performers from a wide array of disciplines, organizations, and backgrounds—and to keep them intensely focused—is to set a finite term for a project and staff it with people working under contracts that last only as long as the jobs they perform contribute to the overall goal." When the finite-term approach was pioneered in the Google Advanced Technology and Projects (ATAP) group, it enabled the contracting of forty computer-vision experts from five different countries for a project that lasted fewer than six months. "We are convinced that we would not have been able to hire even a small fraction of them as permanent employees. Even if we could have, it would have taken more than a year to recruit them and get them working." Dugan and Gabriel, "'Special Forces Innovation.'"

8. Obama, "Executive Order 13526."

9. Chesbrough, "Open Innovation and the Design of Innovation Work."

10. U.S. Department of Defense, "DODTechipedia."

11. U.S. Department of Defense, "DODTechipedia."

12. U.S. Department of Defense, "DODTechipedia."

13. Defense Intelligence Agency, "NeedipeDIA."

14. Priest and Arkin, "Top Secret America."

15. Calandrelli, "An Evaluation of Short Innovation Contest Implementation."

16. Government Accountability Office, "Defense Contracting: Actions Needed to Increase Competition."

17. Government Accountability Office, "Contracting Data Analysis."

18. Orszag, "Improving Government Acquisition."

19. Government Accountability Office, "Federal Contracting: OMB's Acquisition Savings Initiative Had Results." The GAO notes, however, that its estimate was hampered in part by unreliable data. Further, "The potential of this [OMB 2009] initiative was hampered at the start with agencies' general sense of confusion about the savings targets themselves and unclear guidance in a number of areas" (Government Accountability Office, "Defense Contracting: Actions Needed to Increase Competition").

20. Government Accountability Office, "Defense Contracting: Improved Policies and Tools Could Help Increase Competition."

21. Government Accountability Office, "Defense Contracting: Improved Policies and Tools Could Help Increase Competition."

22. Government Accountability Office, "Defense Contracting: Improved Policies and Tools Could Help Increase Competition."

23. Government Accountability Office, "Defense Contracting: Improved Policies and Tools Could Help Increase Competition."

24. Government Accountability Office, "Defense Contracting: Improved Policies and Tools Could Help Increase Competition."

25. The order cited 50 USC § 45b, a statute pertaining to photographing defensive installations (repealed in 1948).

26. Truman, "Executive Order 10290."

27. Gidiere, *The Federal Information Manual.*

28. Elsea, "The Protection of Classified Information"; Pike, "Selected Executive Orders on National Security."

29. Obama, "Executive Order 13526."

30. Oliver, "Last Week Tonight with John Oliver."

31. Elsea, "The Protection of Classified Information."

32. Government Secrecy Reform Act of 1999 (1999). S. 712: 105th Cong.

33. The Supreme Court's decision in *Egan* has been stretched to affirm the president's broad scope of authority to classify information. Department of the Navy v. Egan, 484 U.S. 518, 527 (1988); see Fisher, "Judicial Interpretations of Egan." But that portion of the *Egan* opinion is dicta—not a formal opinion—and does not have judicial precedential value. Further, the Supreme Court in *Egan* also acknowledged the authority of the legislative branch to influence the scope of the executive branch's authority. In fact the judicial branch has not specifically ruled on the scope of the executive branch's implied power to determine whether secrecy is required for national security or to define policies and guidance regarding same.

34. Knezo, "'Sensitive but Unclassified' Information and Other Controls."

35. Public Citizen, "Analysis of Executive Order 13292."

36. Bush, "Title 3—Executive Order 13292 of March 25, 2003."

37. Moynihan and Combest, "Report of the Commission on Protecting and Reducing Government Secrecy."

38. "Classified National Security Information."

39. Reporters Committee for Freedom of the Press, "Statutory Exemption."

40. Public Citizen, "Analysis of Executive Order 13292."

41. Public Citizen, "Analysis of Executive Order 13292."

42. Public Citizen, "Analysis of Executive Order 13292."

43. The CIA claims this authority per the Central Intelligence Agency Act of 1949 (50 US.C., § 403g). For example, see Central Intelligence Agency, "[Redacted] ISCAP Number 2008-020."

44. DOD, Acquisition Law Advisory Panel, *Streamlining Defense Acquisition Laws*, 5–67.

45. 22 CFR § 120.10 Technical data, (3).

46. DOD, Acquisition Law Advisory Panel, *Streamlining Defense Acquisition Laws*, 5–70.

47. DOD, Acquisition Law Advisory Panel, *Streamlining Defense Acquisition Laws*, 5–71.

48. DOD, Acquisition Law Advisory Panel, *Streamlining Defense Acquisition Laws*, 5–71.

49. Invention Secrecy Act, 35 USC § 181-88.

50. DOD, Acquisition Law Advisory Panel, *Streamlining Defense Acquisition Laws*, 5–71

51. 5 USC § 552.

52. U.S. Department of Justice, "FOIA Guide."

53. In FAA v. Robertson, 422 U.S. 255 (1975).

54. U.S. Department of Justice, "FOIA Guide."

55. U.S. Department of Justice, "FOIA Guide."

6. Incentives for Innovation

1. The government *is* able to incentivize employees at federal agencies directly and has defined these incentives in 15 USC § 3710b—Rewards for Scientific, Engineering, and Technical Personnel of Federal Agencies.

2. DOD Acquisition Law Advisory Panel, *Streamlining Defense Acquisition Laws*, 5-1.

3. Maune, "Patent Secrecy Orders."

4. T. Wyatt, "In Search of Reasonable Compensation," 22.

5. T. Wyatt, "In Search of Reasonable Compensation," 2.

6. T. Wyatt, "In Search of Reasonable Compensation," 6.

7. *Cramp & Sons v. International Curtis Marine Turbine Co.*, 246 U.S. 28 (1918)—a case involving a patent holder suing a government contractor for infringement in the course of executing a shipbuilding contract with the U.S. government.

8. The U.S. Supreme Court stated in its opinion, "Under proposals submitted by the Navy Department, the petitioner, the Cramp Company, in 1908 contracted to build two torpedo boat destroyers, Nos. 30 and 31, and in 1911 further contracted to build four such boats, Nos. 47, 48, 49 and 50." *Cramp & Sons v. International Curtis Marine Turbine Co.*, 246 U.S. 28 (1918).

9. An excerpt of FDR's 1918 letter is included in the opinion of Richmond Screw Anchor Co., Inc., v. United States, 275 U.S. 331 (1928)—a case involving a patent holder suing the U.S. government after it contracted with a separate company to construct and install infringing technology; U.S. Government Accountability Office, "B-159356."

10. DOD Acquisition Law Advisory Panel, *Streamlining Defense Acquisition Laws*, 5-1.

11. T. Wyatt, "In Search of Reasonable Compensation," 2.

12. T. Wyatt, "In Search of Reasonable Compensation," 2–3.

13. Brown, "Improving IP Protection for Small Businesses."

14. Quoted in Gaughran, "Whose Idea Was This, Anyway?"

15. Quoted in T. Wyatt, "In Search of Reasonable Compensation."

16. DOD Acquisition Law Advisory Panel, *Streamlining Defense Acquisition Laws*, 5-1.

17. DOD Acquisition Law Advisory Panel, *Streamlining Defense Acquisition Laws*, 5-1.

18. Government Accountability Office: "Contract Management," and "Federal Contracting: Commercial Item Test Program Beneficial."

19. FAR Clause 52.227–1.

20. FAR Clause 52.227–3; Stouck, "Right and Wrong Ways to Use Others' Patents."

21. FAR Clause 52.227–5.

22. T. Wyatt, "In Search of Reasonable Compensation," 8.

23. T. Wyatt, "In Search of Reasonable Compensation," 8.

24. Patent holders may argue that the selective application of the Authorization and Consent clause is unfair or not uniform, but it is unlikely that the Federal Circuit, or even the Court of Federal Claims, would find the application of Authorization and Consent to be arbitrary because the choice of when and which Authorization and Consent clauses to use is a matter committed to agency discretion. Congress did not prohibit the creation of variants of the clause. The government is often afforded the benefit of the doubt in matters where it has discretion, and therefore seemingly unfair application may be excused as long as there is a "rational basis" for the agency's action.

Congress is often hesitant to provide specific guidance to agencies on the application of discretion under the law. Typically the guidance to act in the public interest is considered sufficient. In some cases Congress may specify that agencies should develop rules on the application of laws. The establishment of these rules, called rule-making, is often a cumbersome process, and it must adhere to the procedure specified in the Administrative Procedure Act, 5 USC Subchapter II. Though cumbersome, Congress should consider supplementing the Authorization and Consent law with such a rule-making mandate to provide more uniformity and guidance for agencies applying Authorization and Consent.

25. While this topic is often discussed as a type of "takings" issue, the U.S. Court of Appeals for the Federal Circuit has said that "patent infringement by the government was not a taking under the Fifth Amendment because patent rights, unlike other property rights, are created only by federal statute" (quoted in T. Wyatt, "In Search of Reasonable Compensation"). It is unclear how this squares with Art. I, Sec. 8 of the U.S. Constitution, which requires Congress to create some type of intellectual property rights, or with Justice Taft's rationale in an earlier Supreme Court case of Richmond Screw Anchor Co. v. U.S., 275 U.S. 331 (1928). Regardless of whether such payment is required by the Constitution, however, the issues remain around the statutory requirement that some reasonable compensation be paid and whether that payment is sufficient to incentivize innovation.

26. Robishaw Engineering v. United States, 891 F. Supp. 1134 (E.D. Va. 1995); T. Wyatt, "In Search of Reasonable Compensation."

27. T. Wyatt, "In Search of Reasonable Compensation."

28. Gargoyles, Inc. v. United States, 113 F.3d 1572 (Fed. Cir. 1997)—a case involving a patent holder suing the U.S. government after it purchased infringing eyewear from an alternate manufacturer.

29. T. Wyatt, "In Search of Reasonable Compensation."

30. Quoted in Gaughran, "Whose Idea Was This, Anyway?"

31. An IP Assertions Table is required by DFARS 252,227-7013(e).

32. Fleurant and Perlo-Freeman, "SIPRI Top 100"; Fryer-Biggs and Weisgerber, "U.S. Giants Skimp on Research, Development"; Raytheon, "Form 10-Q."

33. Night Vision Corp. v. U.S., Case No. 06-5048 (Fed. Cir. 2006); Night Vision Corp. v. U.S., 469 F 3d, 1369 (Fed. Cir. 2006); Night Vision Corp. v. U.S., United States Court of Federal Claims, November 8, 2005, No. 03-1214C (Fed. Cl., November 8, 2005).

34. 15 USC § 638.

35. Government Accountability Office, "Military Acquisitions."

36. DARPA Tactical Technology Office, "FANG"; see also DARPA Tactical Technology Office, "Manufacturing Experimentation and Outreach."

37. Innocentive, "Multi-Scale Modeling of Tablet Dissolution" (emphasis added).

38. Author interview with Nathan Wiedenman, DARPA FANG program manager, 2014.

39. The recent 2017 legislation on prize competitions, 15 USC § 3719, provides only high-level guidance on IP policy. It is yet to be seen whether agencies will write new IP regulations or whether government program managers will be restricted to existing FAR/DFARS regulations.

40. Zoltek Corporation v. United States, 672 F.3d 1309, 1311 (Fed.Cir.2012).

41. The *Zoltek* case has bounced back and forth several times between the Court of Federal Claims and the Federal Circuit from its initiation in 1996 through today. Recently a Federal Circuit ruling of February 19, 2016, reversed rulings by the Court of Federal Claims and sent the case back to the latter court, where it is (as of this writing) ongoing. Obviously all this back and forth is very expensive for Zoltek and everyone else involved.

42. T. Wyatt, "In Search of Reasonable Compensation," 17.

43. T. Wyatt, "In Search of Reasonable Compensation."

44. Larson, in "Yesterday's Technology, Tomorrow," pointed to 15 USC § 3710c, which states that government-employed inventors should get 15 percent royalties for licensing their inventions to the government.

45. T. Wyatt, "In Search of Reasonable Compensation."

46. Suh and de Weck, "Modeling Prize-Based Open Design Challenges."

BIBLIOGRAPHY

Abbott, Andrew. *Methods of Discovery: Heuristics for the Social Sciences*. New York: W. W. Norton, 2004.

Ackerman, Evan. "DARPA Robotics Challenge Update," IEEE Spectrum. October 16, 2012. http://spectrum.ieee.org/automaton/robotics/humanoids/iros-2012-darpa-robotics-challenge-update.

Aerospace Industries Association. "Competing for Space: Satellite Export Policy and U.S. National Security." January 2012. Accessed 2015. http://www.aia-aerospace.org/assets/CompetingForSpaceReport.pdf.

Aftergood, Steven. "A Critical Look at Navy v. Egan." Federation of American Scientists. November 16, 2009. Accessed 2014. http://fas.org/blogs/secrecy/2009/11/navy_v_egan/.

———. 2015. "Invention Secrecy Statistics." Federation of American Scientists. Accessed 2015. http://fas.org/sgp/othergov/invention/stats.html.

"After Successful Design Challenge Competition and Testing, DARPA Begins Early Transition of Adaptive Vehicle Make Technologies." February 5, 2014. Accessed May 2018. https://www.darpa.mil/news-events/2014-02-05.

Aitken, Roger. "U.S. Card Fraud Losses Could Exceed $12b by 2020." *Forbes*. October 26, 2016. https://www.forbes.com/sites/rogeraitken/2016/10/26/us-card-fraud-losses-could-exceed-12bn-by-2020/#4f457301d243.

Aitoro, Jill R. "Small Businesses Often Lose in GSA Schedule Contracts, Hill Investigation Finds." May 16, 2013. Accessed March 2015. http://www.bizjournals.com/washington/blog/fedbiz_daily/2013/05/gsa-schedule-contracts-are-often-a.html.

Alexy, Oliver. "Case #2—IBM Going 'Open.'" Open Innovation Community. May 9, 2011. Accessed 2014. http://www.openinnovation.net/researchers/teaching/case-2-ibm-going-open/.

American Association for the Advancement of Science. "Historical Trends in Federal R&D." June 2016. Accessed March 1, 2018. https://www.aaas.org/page/historical-trends-federal-rd.

Anderson, Mike. "What Is the Difference in the DCAA and the DCMA?" November 18, 2013. Accessed 2014. http://www.reliascent.com/government-contracting-blog/bid/102482/What-is-the-difference-in-the-DCAA-and-the-DCMA.

Ansari X-Prize. "Mojave Aerospace Ventures Wins the Competition That Started It All and Launched a Brand New $2 Billion Private Space Industry." N.d. Accessed November 28, 2017. https://ansari.xprize.org/teams.

Arendt, Michael, and Ryan Novak. 2016. *From Incentive Prize and Challenge Competitions to Procurement.* McLean VA: MITRE. Accessed March 1, 2018. https://www.mitre.org/sites/default/files/publications/16-3025-from-incentive-prize-and-challenge-competitions-to-procurement.pdf.

Atherton, Kelsey D. "New Details Emerge on the Surveillance Technology Used to Hunt Osama bin Laden." *Popular Science*. August 30, 2013. Accessed July 2014. http://www.popsci.com/technology/article/2013-08/nsa-satellites-hunted-bin-laden.

Avionics Department. "NAWCWD TP 8347." October 2013. Accessed August 2014. www.navair.navy.mil/nawcwd/ewssa/downloads/NAWCWD%20TP%208347.pdf.

Axe, David. "Did the Marines' 40-Year-Old 'Amphibious Tractor' Just Strike Again?" September 14, 2017. Accessed May 2018. https://www.thedailybeast.com/did-the-marines-40-year-old-amphibious-tractor-just-strike-again.

Badani, Jayesh. "Metrics for Open Innovation—What's My Open Innovation Quotient." May 2009. Accessed December 2014. http://blog.ideaken.com/2009/05/metrics-for-open-innovation-whats-my.html.

Bacon, Lance M. "Inside the Amphibious Vehicles That Won the Marines' $225m Contracts," *Marine Corps Times*, January 4, 2016. Accessed 2018. https://www.marinecorpstimes.com/news/your-marine-corps/2016/01/04/inside-the-amphibious-vehicles-that-won-the-marines-225m-contracts.

Baldwin, Carliss Y., and Joachim Henkel. *The Impact of Modularity on Intellectual Property and Value Appropriation*. Cambridge MA: Harvard Business School, 2011.

———. "Modularity and Intellectual Property Protection." *Strategic Management Journal* 36, no. 11 (2015): 1637–55.

Barnett, Chance. "Donation-Based Crowdfunding Sites: Kickstarter Vs. Indiegogo." *Forbes.* September 9, 2013. Accessed 2014. http://www.forbes.com/sites/chancebarnett/2013/09/09/donation-based-crowdfunding-sites-kickstarter-vs-indiegogo/.

Birkinshaw, Julian, and Jules Goddard. "Combine Harvesting." *London Business School Labnotes*. 2009. https://www.innocentive.com/files/node/casestudy/roche-experience-open-innovation.pdf.

Bonadio, Steve. "How to Accelerate Innovation through Challenge Driven Innovation." InnovationManagement.se. 2011. Accessed December 2014. http://www.innovationmanagement.se/2011/10/17/how-to-accelerate-innovation-through-challenge-driven-innovation/.

Boyle, Rebecca. "How the First Crowdsourced Military Vehicle Can Remake the Future of Defense Manufacturing." *Popular Science*. June 30, 2011. Accessed February 2015. http://www.popsci.com/cars/article/2011-06/how-first-crowdsourced-military-car-can-remake-future-defense-manufacturing.

Bradley, Stephanie. "Late Loaning Lenders: Bringing Awareness to Expiring Loans." Kiva. February 2, 2013. Accessed May 2014. http://blog.kiva.org/tags/expirations.

Branscomb, Lewis M., and James H. Keller. *Investing in Innovation: Creating a Research and Innovation Policy That Works.* Cambridge MA: MIT Press, 1999.

Brinkley, Douglas. "50 Years Ago, Kennedy Reached for Stars in Historic Rice Address." *Houston Chronicle*. September 10, 2012. accessed 2014. http://www.chron.com/news/nation-world/article/50-years-ago-Kennedy-reached-for-stars-in-3852085.php.

Brown, Alison. "Improving IP Protection for Small Businesses." NDIA. DTIC. 2009. Accessed January 23, 2018. www.dtic.mil/get-tr-doc/pdf?AD=AD1008934.

Bush, George W. "Title 3—Executive Order 13292 of March 25, 2003." White House. March 25, 2003. Accessed 2014. http://fas.org/sgp/bush/eo13292inout.html.

Bushell, Edward, and Peter Goon. "The Unintended Consequences of the Defence Trade Controls Act 2012." Submission to Inquiry into the Implementation of the Defence Trade Controls Legislation, Defence and Trade Committee. 2013. Accessed November 7, 2017. https://www.aph.gov.au/DocumentStore.ashx?id=0df8d226-bef4-415d-a435-68468628c465&subId=23369.

Calandrelli, Emily. "An Evaluation of Short Innovation Contest Implementation in the Federal Context." PhD diss., Massachusetts Institute of Technology, 2013.

Camiña, Steven L. "A Comparison of Taxonomy Generation Techniques Using Bibliometric Methods: Applied to Research Strategy Formulation." PhD diss., Massachusetts Institute of Technology, 2010.

Carter, Ashton B., and John P. White. *Keeping the Edge: Managing Defense for the Future*. Cambridge MA: MIT Press, 2001.

Center for Strategic and International Studies. "Briefing of the Working Group on the Health of the U.S. Space Industrial Base and the Impact of Export Controls." February 2008. Accessed December 6, 2017. https://csis-prod.s3.amazonaws.com/s3fs-public/legacy_files/files/media/csis/pubs/021908_csis_spaceindustryitar_final.pdf.

Central Intelligence Agency. "Careers and Internships." U.S. Central Intelligence Agency. September 21, 2009. Accessed 2014. https://www.cia.gov/careers/opportunities/science-technology/view-jobs.html.

———. "We can neither confirm nor deny that this is our first tweet." June 6, 2014. Accessed 2014. https://twitter.com/cia.

———."The President's Daily Brief." January 15, 1969. Accessed 2014. http://www.archives.gov/declassification/iscap/pdf/2005-014-doc4.pdf.

———. "[Redacted] ISCAP Number 2008-020." November 17, 1999. Accessed 2014. http://www.archives.gov/declassification/iscap/pdf/2008-020-doc1.pdf.

———."The Uraba Massacres." May 9, 1998. Accessed 2014. http://www.archives.gov/declassification/iscap/pdf/2009-064-doc3.pdf.

Chafkin, Max. "The Customer Is the Company." June 1, 2008. Accessed 2014. http://www.inc.com/magazine/20080601/the-customer-is-the-company.html.

Chesbrough, Henry. "Managing Open Innovation." *Research-Technology Management* 47, no. 1 (2004): 23–26. http://cms.sem.tsinghua.edu.cn/semcms/res_base/semcms_com_www/upload/home/store/2008/7/3/2979.pdf.

———. *Open Innovation: The New Imperative for Creating and Profiting from Technology*. Cambridge MA: Harvard Business Press, 2003.

———. "Open Innovation and the Design of Innovation Work." *Forbes*. May 18, 2011. Accessed 2014. http://www.forbes.com/sites/henrychesbrough/2011/05/18/open-innovation-and-the-design-of-innovation-work/.

Chesbrough, Henry, and A. Crowther. "Beyond High Tech: Early Adopters of Open Innovation in Other Industries." *R&D Management* 36, no. 3 (2006): 229–36.

Chu, Jennifer. "Searching for Balloons in a Social Network." MIT. October 28, 2011. Accessed 2015. http://newsoffice.mit.edu/2011/red-balloons-study-102811.

Cisco. "Cisco Announces Winners of the First Cisco Entrepreneurs in Residence (EIR) Program in Europe at the Pioneers Festival, Vienna." Cisco. October 29, 2014. Accessed 2015. http://newsroom.cisco.com/release/1518065/Cisco-Announces-Winners-of-the-First-Cisco-Entrepreneurs-in-Residence-EIR-Program-in-Europe-at-The-Pioneers-Festival-Vienna?utm_medium=rss.

———. "Cisco Entrepreneurs in Residence." CrunchBase. 2014. Accessed 2014. https://www.crunchbase.com/organization/cisco-accelerator-for-entrepreneurs.

———. "Cisco Entrepreneurs in Residence: Innovation Program for Early-Stage Entrepreneurs in Enterprise Web, Cloud & IoT/IoE." AngelList. 2014. Accessed 2014. https://angel.co/cisco-entrepreneurs-in-residence-2.

Cisco Systems, Inc. "Cisco Entrepreneurs in Residence." Cisco. 2004. Accessed 2014. https://eir.cisco.com/.

Clarkson, M. "A Stakeholder Framework for Analyzing and Evaluating Corporate Social Performance." *Academy of Management Review* 20 (1995): 92–117.

"Classified National Security Information." Federal Register, 2014.

Cleland, Brian, Brendan Galbraith, Barry Quinn, and Paul Humphreys. "Platform Strategies for Open Government Innovation." In *Proceedings for the 8th European Conference on Innovation and Entrepreneurship: ECIE 2013*. Academic Conferences Limited, 2013.

Clinton, William Jefferson. "The National Security Science and Technology Strategy." White House. 1995. Accessed July 2014. http://clinton4.nara.gov/WH/EOP/OSTP/nssts/html/letter.html.

CNBC. "Quirky." CNBC. May 14, 2013. Accessed 2014. http://www.cnbc.com/id/100731600#.

Conrad, Abigail, Tulika Narayan, Judy Geyer, Stephen Bell, and Luciano Kay. *A Framework for Evaluating Innovation Challenges*. Bethesda MD: Abt Associates, March 2017.

Cooper, R., and S. Edgett. "Maximizing Productivity in Product Innovation." *Research-Technology Management* 51, no. 2 (2008): 47–58.

Coviello, Nicole E., and Marian V. Jones. "Methodological Issues in International Entrepreneurship Research." *Journal of Business Venturing* 19, no. 4 (2004): 485–508.

Cropley, David H. *Creativity in Engineering: Novel Solutions to Complex Problems*. San Diego: Academic Press, 2015.

CrunchBase. "In-Q-Tel." CrunchBase. 2014. Accessed December 2014. http://www.crunchbase.com/organization/in-q-tel.

Dahlander, L., and D. Gann. "How Open Is Innovation?" *Research Policy* 39, no. 6 (2010): 699–709.

Dalton, R. "A Case Study of the Advanced Amphibious Assault Vehicle (AAAV) Program from a Contracting Perspective." PhD diss., Naval Postgraduate School, 1998. http://www.dtic.mil/dtic/tr/fulltext/u2/a359054.pdf.

DARPA. "After Successful Design Challenge Competition and Testing, DARPA Begins Early Transition of Adaptive Vehicle Make Technologies." DARPA. February 5, 2014. Accessed December 2014. http://www.darpa.mil/NewsEvents/Releases/2014/02/05.aspx.

———. "After Successful Design Challenge Competition and Testing, DARPA Begins Early Transition of Adaptive Vehicle Make Technologies." DARPA. February 5, 2014. Accessed May 2018. https://www.darpa.mil/news-events/2014-02-05.

———. "Component, Context, and Manufacturing Model Library 1 (C2M2L-2) Solicitation Number: DARPA-BAA-11-47. 6 2011." June, 2011. Accessed December 2014. https://www.fbo.gov/index?s=opportunity&mode=form&id=e9bd71694d0ca43332c8d98ca37f4d28&tab=core&_cview=1.

———. "DARPA Today: Anticipating and Meeting New Challenges." October 2, 2014. Accessed 2015. www.darpa.mil/WorkArea/DownloadAsset.aspx?id=2147488464.

———. "DARPA's Shredder Challenge Solved." DARPA. 2011. http://www.darpa.mil/NewsEvents/Relcases/2011/12/02_.aspx.

———. "FAQS—Open Manufacturing." Accessed December 2014. http://www.darpa.mil/workarea/downloadasset.aspx?id=2147483957.

DARPA Information Innovation Office (120). "Cyber Fast Track (CFT) DARPA-RA-11-52." August 3, 2011. Accessed 4 2015. https://www.fbo.gov/index?s=opportunity&mode=form&id=897a16b7d794ffa357de54b48dd5a99d&tab=core&_cview=1.

DARPA Tactical Technology Office. "Component, Context, and Manufacturing Model Library-2." Broad Agency Announcement, DARPA, 2012. Accessed December 7, 2017. https://www.fbo.gov/index?s=opportunity&mode=form&id=de840663fd40f07f6825e72fb97bbfc9&tab=core&_cview=0.

———. "FANG; Fast, Adaptable, Next-Generation Ground Vehicle; DARPA-BAA-12-15." Broad Agency Announcement, DARPA, 2011. https://www.fbo.gov/index?s=opportunity&mode=form&id=502e4d989277e9d294a3ac744b0a5ac6&tab=core&_cview=1.

———. "Manufacturing Experimentation and Outreach (MENTOR)." Broad Agency Announcement, DARPA, 2010.

Davis, L., and J. Davis. "How Effective Are Prizes as Incentives to Innovation? Evidence from Three 20th Century Contests." *DRUID Summer Conference*, 2004.

de Weck, Olivier. "Fast Adaptable Next-Generation Ground Vehicle Challenge, Phase 1 (FANG-1) Post-Challenge Analysis." DARPA. 2013.

de Weck, Olivier, James Lyneis, and Dan Braha. "ESD.36 System Project Management Lecture 1." September 2012. Accessed 2015. http://ocw.mit.edu/courses/engineering-systems-division/esd-36-system-project-management-fall-2012/lecture-notes/MITESD_36F12_Lec01.pdf.

Dean, Josh. 2013. "Is This the World's Most Creative Manufacturer?" Inc. 10. 2013. Accessed 2014. http://www.inc.com/magazine/201310/josh-dean/is-quirky-the-worlds-most-creative-manufacturer.html.

Defense Acquisition University. "Omnibus Contracts." N.d. Accessed August 2014. https://dap.dau.mil/aap/pages/qdetails.aspx?cgiSubjectAreaID=22&cgiQuestionID=26887.

Defense Industry Daily. "BAE Closes United Defense LP Acquisition." June 27, 2005. Accessed December 2014. http://www.defenseindustrydaily.com/bae-closes-united-defense-lp-acquisition-0755/.

Defense Intelligence Agency. "NeedipeDIA." 2014. Accessed December 2014. http://www.dia.mil/Business/Needipedia/Needipedia-FAQ/.

Defense Security Service. "Facility Clearance Process FAQs." U.S. Department of Defense. N.d. Accessed April 2015. http://www.dss.mil/isp/fac_clear/per_sec_clear_proc_faqs.html.

———. "A Guide for the Preparation of a DD Form 254." June 2013. Accessed April 2015. http://www.cdse.edu/documents/cdse/DD254.pdf.

DeLanda, Manuel. *War in the Age of Intelligent Machines*. New York: Zone Books, 1991.

Delaney, Steve. "Securing Net-Centric cots Systems." *COTS Journal*. October 2010. Accessed March 2015. http://www.cotsjournalonline.com/articles/view/101574.

"Dell Expands Its Commitment to Entrepreneurship and Innovation with $300 Million Strategic Innovation Venture Fund." Dell. December 12, 2013. Accessed 2014. http://www.dell.com/learn/us/en/uscorp1/secure/2013-12-12-dell-ventures-entrpreneur-strategic-innovation-fund.

"Dell Ventures." CrunchBase. 2014. Accessed 2014. https://www.crunchbase.com/organization/dell-venture.

Denett, P. "Enhancing Competition in Federal Acquisition." Executive Office of the President, Office of Management and Budget. 2007. http://www.whitehouse.gov/sites/default/files/omb/procurement/comp_contracting/competition_memo_053107.pdf.

Department of Homeland Security, Office of Inspector General. "Reducing Overclassification of DHS' National Security Information." August 2013. Accessed 2014. http://www.oig.dhs.gov/assets/Mgmt/2013/OIG_13-106_Jul13.pdf.

DOD. "DOD SBIR/STTR Program Desk Reference." U.S. Department of Defense. October 7, 2011. Accessed 2014. http://www.acq.osd.mil/osbp/sbir/sb/resources/deskreference/13_phas3.shtml.

DOD, Acquisition Law Advisory Panel. *Streamlining Defense Acquisition Laws*. Washington DC: Defense Systems Management College Press, 1993. doi:AD-A264919.

Donaldson, T., and L. Preston. "The Stakeholder Theory of the Corporation: Concepts, Evidence, and Implications." *Academy of Management Review* 20, no. 1 (1995). http://www.jstor.org/stable/258887.

Doyle, John M. "Outside the Box: DARPA Wants to Help the Marine Corps Build Vehicles Faster and Cheaper." The Corps. April 2012. Accessed February 2015. https://4gwar.files.wordpress.com/2011/03/april-darpa-seapower-april-2012-2.pdf.

Drach-Zahavy, A., and A. Somech. "Understanding Team Innovation: The Role of Team Processes and Structures." *Group Dynamics: Theory, Research, and Practice* 5, no. 2 (2001): 111.

Drew, Christopher. "Attack on Bin Laden Used Stealthy Helicopter That Had Been a Secret." *New York Times*. May 5, 2011. Accessed July 2014. http://www.nytimes.com/2011/05/06/world/asia/06helicopter.html.

Driessnack, John D. "Unique Transaction Costs in Defense Market(s): The Explanatory Power of New Institutional Economics." DTIC document, 2005.

Drummond, K. "Programmers Shred Pentagon's Paper Puzzle Challenge." 2011. http://www.wired.com/2011/12/darpa-shredder-challenge-2/.

Dugan, Regina E., and Kaigham J. Gabriel. 2013. "'Special Forces Innovation': How DARPA Attacks Problems." *Harvard Business Review* 10:74–82.

Dwyer, Morgan, et al. "The Global Impact of ITAR on the For-Profit and Non-Profit Space Communities." International Astronautical Federation, 2012.

Egan, M. "In-Q-Tel: A Glimpse Inside the CIA's Venture-Capital Arm." Fox Business. June 14, 2013. Accessed December 2014. http://www.foxbusiness.com/technology/2013/06/14/in-q-tel-glimpse-inside-cias-venture-capital-arm/.

Eighty-First Congress. 1950. "Public Law 507—National Science Foundation Act of 1950."

Electronic Information Products Division, Patent Technology Monitoring Team. "Description of Patent Types." U.S. Patent and Trademark Office. October 3, 2013. Accessed September 2014. http://www.uspto.gov/web/offices/ac/ido/oeip/taf/patdesc.htm.

"Eligibility Guidelines for Gaining Security Clearance." Military.com. 2015. Accessed 2015. http://www.military.com/veteran-jobs/security-clearance-jobs/security-clearance-eligibility.html.

Elsea, Jennifer K. "The Protection of Classified Information: The Legal Framework." January 10, 2013. Accessed 2014. http://fas.org/sgp/crs/secrecy/RS21900.pdf.

Elwell, Andrew. 2013. "DARPA Offers $1 Million Prize for New Amphibious Armoured Vehicle Designs." Defence iQ. July 8, 2013. Accessed April 2015. http://www.defenceiq.com/amoured-vehicles/articles/darpa-offers-1-million-prize-for-new-amphibious-ar/.

Eremenko, Paul, and Nathan Wiedenman. "Adaptive Vehicle Make (AVM)." Proposers' Day Briefing, DARPA Tactical Technology Office, 2010.

Eriksson Lundström, Jenny, Mikael Wiberg, Stefan Hrastinski, Mats Edenius, and Pär J. Ågerfalk. *Managing Open Innovation Technologies.* Spring Science and Business Media, 2013.

Federal Service Desk. "What Is a CAGE Code? How Are CAGE Codes Assigned?" Federal Service Desk. Accessed 2014. https://www.fsd.gov/fsd-gov/answer.do?sysparm_number=KB0011119.

Feickert, Andrew. "Marine Corps Amphibious Combat Vehicle (ACV) and Marine Personnel Carrier (MPC): Background and Issues for Congress." R42723, Congressional Research Service. 2017. Accessed January 29, 2018. https://fas.org/sgp/crs/weapons/R42723.pdf.

Feng, W., E. Crawley, O. de Weck, D. Lessard, and B. Cameron. "Understanding the Impacts of Indirect Stakeholder Relationships—Stakeholder Value Network Analysis and Its Application to Large Engineering Projects." *Strategic Management Society* (2012-AC-1344). https://www.federalregister.gov/articles/2014/07/30/2014-17836/classified-national-security-information.

Fisher, Louis. "Judicial Interpretations of Egan." November 13, 2009. Accessed 2014. http://fas.org/sgp/eprint/egan.pdf.

Fleurant, Aude, and Sam Perlo-Freeman. "SIPRI Top 100 and Recent Trends in the Arms Industry." Stockholm International Peace Research Institute. December 2014. Accessed 2015. http://www.sipri.org/research/armaments/production/recent-trends-in-arms-industry.

"Fold It: Solve Puzzles for Science." Foldit. January 28, 2018. Accessed January 28, 2018. https://fold.it/.

Fossum, Donna, Lawrence S. Painter, Valerie L. Williams, Allison Yezril, and Elaine M. Newton. "Discovery and Innovation: Federal Research and Development Activities in the Fifty States, District of Columbia, and Puerto Rico." 2000. Accessed 2014. http://www.rand.org/pubs/monograph_reports/MR1194.html.

Freedberg, S. "Amos Says Marines to Drop High Speed ACV, for Now; Phased Approach Likely. Breaking Defense. January 29, 2014. Accessed December 2014. http://breakingdefense.com/2014/01/amos-says-marines-to-drop-high-speed-acv-phased-approach-likely/.

———. "Marines 2014: Year of Decision for Amphibious Combat Vehicle." Breaking Defense. January 9, 2014. Accessed December 2014. http://breakingdefense.com/2014/01/marines-2014-year-of-decision-for-amphibious-combat-vehicle/.

Freeman, R. E. *Strategic Management: A Stakeholder Approach.* Boston: Pitman, 1984.

Frey, K., L. Christian, and S. Haag. "Whom Should Firms Attract to Open Innovation Platforms? The Role of Knowledge Diversity and Motivation." *Long Range Planning* 44, no. 5 (2011): 397–420.

Fryer-Biggs, Zachary, and Marcus Weisgerber. "U.S. Giants Skimp on Research, Development." Defense News. August 19, 2013. Accessed 2015. http://archive.defensenews.com/article/20130819/DEFREG02/308190005/US-Giants-Skimp-Research-Development.

Gassman, O., E. Enkel, and H. Chesbrough. "The Future of Open Innovation." *R&D Management* 40, no. 3 (2010): 213–21.
Gaughran, Madeleine B. "Whose Idea Was This, Anyway? When the Pentagon Is the Buyer, Your Invention Belongs to Everyone." March 19, 2001. https://www.wsj.com/articles/SB984598977619355038.
General Dynamics. "General Dynamics to Continue Amphibious Combat Vehicle Testing for Marine Corps." July 11, 2014. Accessed January 29, 2018. https://www.gd.com/news/press-releases/2014/07/general-dynamics-continue-amphibious-combat-vehicle-testing-marine-corps.
General Electric. "FirstBuild." General Electric. N.d. Accessed 2014. https://firstbuild.com/.
———. 2015. "GE Open Innovation." General Electric. 2015. Accessed 2015. http://www.ge.com/about-us/openinnovation.
General Services Administration. "Award M6785401C0001." March 16, 2007. Accessed December 2014. https://www.fpds.gov/ezsearch/fpdsportal?indexName=awardfull&templateName=1.4.2&s=FPDSNG.COM&q=General+Dynamics+Amphibious+Sy+PIID%3A%22M6785401C0001%22+P00291&x=17&y=14.
Gidiere, P. Stephen. *The Federal Information Manual: How the Government Collects, Manages, and Discloses Information under FOIA and Other Statutes.* N.p.: American Bar Association, 2006.
Glaser, April. "Biometrics Are Coming, Along with Serious Security Concerns." Wired. March 9, 2016. https://www.wired.com/2016/03/biometrics-coming-along-serious-security-concerns/.
Golightly, J., C. Ford, P. Sureka, and B. Reid. "Realising the Value of Open Innovation." Big Innovation Centre. 2012. http://biginnovationcentre.com/Assets/Docs/Reports/Realising_theValue_ofOI_FINAL.pdf.
Gompers, P., and J. Lerner. *The Venture Capital Cycle*. Cambridge MA: MIT Press, 2004.
Gourley, B. "DARPA's Cyber Fast Track Adds Agility to Research Funding." GovLoop. November 24, 2011. Accessed December 2014. https://www.govloop.com/community/blog/darpas-cyber-fast-track-adds-agility-to-research-funding/.
Government Accountability Office. "Amphibious Combat Vehicle: Some Acquisition Activities Demonstrate Best Practices; Attainment of Amphibious Capability to Be Determined." Government Accountability Office. 2015. https://www.gao.gov/assets/680/673405.pdf.
———. "B-159356." 46 COMP. GEN. 227. Government Accountability Office. 1966. Accessed January 5, 2018. https://www.gao.gov/products/B-159356#mt=e-report.
———. "Contract Management: Benefits of Simplified Acquisition Test Procedures Not Clearly Demonstrated." Government Accountability Office. 2001. http://www.gao.gov/new.items/d01517.pdf.
———. "Contracting Data Analysis: Assessment of Government-Wide Trends." GAO-17-244SP. Government Accountability Office. 2017. https://www.gao.gov/assets/690/683273.pdf.

———. "Defense Contracting: Actions Needed to Increase Competition." Government Accountability Office. 2013. http://www.gao.gov/assets/660/653404.pdf.
———. "Defense Contracting: Improved Policies and Tools Could Help Increase Competition on DOD's National Security Exception Procurements." Government Accountability Office. 2012. http://www.gao.gov/assets/590/587681.pdf.
———. "Federal Contracting: Commercial Item Test Program Beneficial, but Actions Needed to Mitigate Potential Risks." Government Accountability Office. 2014. http://www.gao.gov/assets/670/660650.pdf.
———. "Federal Contracting: OMB's Acquisition Savings Initiative Had Results, but Improvements Needed." Government Accountability Office. 2011. http://www.gao.gov/assets/590/586265.pdf.
———. "Military Acquisitions: DOD Is Taking Steps to Address Challenges Faced by Certain Companies." Government Accountability Office. July 2017. https://www.gao.gov/assets/690/686012.pdf.
———. "Open Innovation: Executive Branch Developed Resources to Support Implementation, but Guidance Could Better Reflect Leading Practices." GAO-17-507. Government Accountability Office. June 2017. Accessed March 1, 2018. https://www.gao.gov/assets/690/685161.pdf.
Government Secrecy Reform Act of 1999. S. 712. 105th Cong.
Grant, R. "Kiva Gets $3m Award from Google to Reach the 'Overlooked' Poor with Big Ideas." VentureBeat. 2013. http://venturebeat.com/2013/12/12/kiva-gets-3m-award-from-google-to-reach-the-overlooked-poor-with-big-ideas/.
Grasso, Valerie. "Defense Acquisition: Use of Lead System Integrators (LSIS)—Background, Oversight Issues, and Options for Congress." Congressional Research Service. 2010. http://www.fas.org/sgp/crs/natsec/RS22631.pdf.
Grimshaw, Miles. "Analysis of Quirky: Do Consumers Know What They Want?" June 11, 2014. Accessed 2014. http://milesgrimshaw.com/analysis-quirky/.
Grossi, I. "Stakeholder Analysis in the Context of the Lean Enterprise." Master's thesis, Massachusetts Institute of Technology, 2003.
Hamilton, A. "50 Best Websites 2008." *Time Magazine*. 2008. http://content.time.com/time/specials/2007/article/0,28804,1809858_1809952_1811306,00.html.
Harrigan, Kathryn Rudie. "Research Methodologies for Contingency Approaches to Business Strategy." *Academy of Management Review* 8, no. 3 (1983): 398–405.
Hauser, J., G. Tellis, and A. Griffin. "Research on Innovation: A Review and Agenda for Marketing Science." *Marketing Science* 25, no. 6 (2006): 687–717.
Henkel, Joachim. "Selective Revealing in Open Innovation Processes: The Case of Embedded Linux." *Research Policy* 35, no. 7 (2006): 953–69.
Henn, S. "In-Q-Tel: The CIA's Tax-Funded Player in Silicon Valley." National Public Radio. 2012. http://www.npr.org/blogs/alltechconsidered/2012/07/16/156839153/in-q-tel-the-cias-tax-funded-player-in-silicon-valley.
Heredero, Carmen de Pablos. *Open Innovation in Firms and Public Administrations: Technologies for Value Creation.* N.p.: IGI Global, 2011.

"How Quirky Turns Ideas into Inventions." *Popular Mechanics*. January 7, 2014. Accessed March 2015. http://www.popularmechanics.com/technology/gadgets/a9946/how-quirky-turns-ideas-into-inventions-16344763/

Howells, John. *The Management of Innovation and Technology*. London: Sage, 2005.

INCOSE (International Council on Systems Engineering). "Systems Engineer: Guru or Scientist?" 2014. Accessed August 2014. http://www.incose.org/symp2014/?page=techprogram&type=industry&id=academic.

———. "What Is Systems Engineering." INCOSE. N.d. Accessed March 2015. http://www.incose.org/practice/whatissystemseng.aspx.

Indiegogo. "Fees & Pricing." Indiegogo. 2015. Accessed 2015. https://support.indiegogo.com/hc/en-us/articles/204456408-Fees-Pricing.

"Industrial Research Institute." Industrial Research Institute. 2014. Accessed July 2014. www.iriweb.org.

Info Security. "DARPA Says Goodbye to Hacker-Friendly Cyber Fast Track Program." March 7, 2013. Accessed March 2013. http://www.infosecurity-magazine.com/news/darpa-says-goodbye-to-hacker-friendly-cyber-fast/.

Information Security Oversight Office. "2013 Report to the President." 2013. Accessed 2014. http://www.archives.gov/isoo/reports/2013-annual-report.pdf.

Innocentive. "Harvesting the Energy in Buildings." 2014. Accessed December 2014. https://www.innocentive.com/ar/challenge/9933685.

———. "Multi-Scale Modeling of Tablet Dissolution." November 3,2014. Accessed March 1, 2018. https://www.innocentive.com/ar/challenge/9933558.

"Innovation Awards: Our Annual Prizes Recognise Successful Innovators in Eight Categories." *The Economist*. 2011. http://www.economist.com/node/21540389.

Interagency Security Classification Appeals Panel. "NARA and Declassification." August 25, 2014. Accessed October 2014. http://www.archives.gov/declassification/iscap/decision-table.html.

Jacobsen, Robert. "Technology Readiness Levels Introduction." NASA. October 21, 2004. Accessed September 2014. http://web.archive.org/web/20051206035043/http://as.nasa.gov/aboutus/trl-introduction.html.

Jenkins, Nash. "Feds: Russian Hackers Are Attacking U.S. Power Plants." March 16, 2018. Accessed May 2018. http://time.com/5202774/russia-hacking-dhs-report-power/.

Johnson, Brian T. "The New Space Race: Challenges for U.S. National Security and Free Enterprise." Heritage Foundation. August 25, 1999. Accessed 2014. http://www.heritage.org/research/reports/1999/08/the-new-space-race.

Jones, P. "FAR Letter to National Defense Magazine." Tenebraex Corporation. N.d. http://www.ndia.org/Divisions/Divisions/SmallBusiness/Documents/Content/ContentGroups/Divisions1/Small_Business/PDFS12/FAR%20Letter%20to%20National%20Defense%20Magazine.pdf.

Judson, Jen. "BAE Wins Marin Corps Contract to Build New Amphibious Combat Vehicle," *Defense News*, June 19, 2018. Accessed 2018. https://www

.defensenews.com/land/2018/06/19/bae-wins-marine-corps-contract-to-build-new-amphibious-combat-vehicle.

Kanani, R. "Incentivizing Innovation: How the White House Uses Challenge.gov to Solve Big Problems." *Forbes*. 2014. http://www.forbes.com/sites/rahimkanani/2014/02/17/incentivizing-innovation-how-the-white-house-uses-challenge-gov-to-solve-big-problems/.

Kendall, F. "Public Access to the Results of Department of Defense–Funded Research." Washington DC: Department of Defense Acquisition, Technology, and Logistics. 2014. http://dtic.mil/dtic/pdf/PublicAccessMemo2014.pdf.

Kennedy, John Fitzgerald. "John F. Kennedy Moon Speech—Rice Stadium." Rice University. N.d. Accessed 2014. http://er.jsc.nasa.gov/seh/ricetalk.htm.

Kickstarter. "Creator Questions." Kickstarter. 2015. Accessed 2015. https://www.kickstarter.com/help/faq/creator+questions.

Knezo, Genevieve J. "'Sensitive but Unclassified' Information and Other Controls: Policies and Options for Scientific and Technical Information." December 29, 2006. Accessed 2014. http://fas.org/sgp/crs/secrecy/RL33303.pdf.

Kokalitcheva, Kia. "For Google Ventures, 2014 Yielded 16 Exits and a Strong Focus on Life Sciences and Health Tech." VentureBeat. December 15, 2014. Accessed 2015. http://venturebeat.com/2014/12/15/for-google-ventures-2014-yielded-16-exits-and-a-strong-focus-on-life-sciences-and-health-tech/.

Kolodny, Lora. "Indiegogo Raises $40m in Largest Venture Investment Yet for Crowdfunding Startup." *Wall Street Journal*. January 28, 2014. Accessed 2014. http://blogs.wsj.com/venturecapital/2014/01/28/indiegogo-raises-40m-in-largest-venture-investment-yet-for-crowdfunding-startup/.

Lakhani, Karim R., David A. Garvin, and Eric Longstein. "TopCoder (A): Developing Software through Crowdsourcing." January 15, 2010. Accessed 2014. https://hbr.org/product/TopCoder—A—Developing-/an/610032-PDF-ENG.

Lakhani, Karim R., Lars Bo Jeppesen, Peter Andreas Lohse, and Jill A. Panetta. *The Value of Openess in Scientific Problem Solving*. Cambridge MA: Division of Research, Harvard Business School, 2007.

Lapray, M., and S. Rebouillat. "Innovation Review: Closed, Open, Collaborative, Disruptive, Inclusive, Nested . . . and Soon Reverse. How about the Metrics: Dream and Reality." *International Journal of Innovation and Applied Studies* 9, no. 1 (2014).

Larson, Daniel. "Yesterday's Technology, Tomorrow: How the Government's Treatment of Intellectual Property Prevents Soldiers from Receiving the Best Tools to Complete Their Mission." *John Marshall Review of Intellectual Property Law* (2007): 171. https://heinonline.org/HOL/LandingPage?handle=hein.journals/johnmars7&div=10&id=&page.

Lee, Sang M., Taewon Hwang, and Donghyun Choi. "Open Innovation in the Public Sector of Leading Countries." *Management Decision* 50, no. 1 (2012): 147–62.

"Legal Definition of Dicta." 'Lectric Law Library. 2015. Accessed 2015. http://www.lectlaw.com/def/d047.htm?PageSpeed=noscript.

Leonard, Scott, and Mel Hafer. "Advanced Manufacturing Enterprise Strategic Baseline." July 15, 2011. Accessed August 2014. https://www.dodmantech.com/JDMTP/Files/AME_Strategic_Baseline_03_Nov_11.pdf.
Lexington Institute. "Dragons of Change: The U.S. Marine Corps' Advanced Amphibious Assault Vehicles." *Armed Forces Journal International*. 1999. Accessed January 3, 2018. http://www.lexingtoninstitute.org/dragons-of-changethe-u-s-marine-corps-advanced-amphibious-assault-vehicles/.
Li, William, Pablo Azar, David Larochelle, Phil Hill, and Andrew W. Lo. "Law Is Code: A Software Engineering Approach to Analyzing the United States Code." *Journal of Business and Technology Law* 10, no. 2 (2015).
Lim, D. "DARPA's New 'Fast Track' Okays Hacker Projects in Just Seven Days." *Wired Magazine*. November 14, 2011. http://www.wired.com/2011/11/darpa-fast-track/.
Ling, Brian. "User Centered Innovation Is Dead." Design Sojourn. February 5, 2010. Accessed 2014. http://www.designsojourn.com/user-centered-innovation-is-dead/.
Loudin, Kathlyn Hopkins. "Lead Systems Integrators: A Post-Acqusition Reform Retrospective." Defense Acqusition University. January 2010. Accessed April 2015. http://www.dau.mil/pubscats/pubscats/ar%20journal/arj53/loudin53.pdf.
Love, Dylan. "How Lego Uses the Internet to Turn Your Creations into Amazing Products." Business Insider. June 23, 2014. Accessed 2014. http://www.businessinsider.com/lego-ideas-2014-6.
Ludwig, Adam. "Don't Call It Crowdsourcing: Quirky CEO Ben Kaufman Brings Invention to the Masses." *Forbes*. April 23, 2012. Accessed 2014. http://www.forbes.com/sites/techonomy/2012/04/23/dont-call-it-crowdsourcing-quirky-ceo-ben-kaufman-brings-invention-to-the-masses/.
Manufacturing USA. "Annual Report." Manufacturing USA. 2016. https://www.manufacturingusa.com/sites/prod/files/Manufacturing%20USA-Annual%20Report-FY%202016-web.pdf.
Marshall, J. "Online Gamers Achieve First Crowd-Sourced Redesign of Protein." *Scientific American*. January 22, 2012. http://www.scientificamerican.com/article/victory-for-crowdsourced-biomolecule2/.
Massey, Kevin. "Adaptive Vehicle Make (AVM)." DARPA Tactical Technology Office. N.d. Accessed April 2015. http://www.darpa.mil/Our_Work/TTO/Programs/Adaptive_Vehicle_Make__%28AVM%29.aspx.
Maune, James. "Patent Secrecy Orders: Fairness Issues in Application of Invention of Secrecy Act." *Texas Intellectual Property Law Journal* 20. http://www.tiplj.org/wp-content/uploads/Volumes/v20/v20p471.pdf.
McCain, John. "Restoring American Power." 2017. Accessed February 24, 2018. https://www.mccain.senate.gov/public/_cache/files/25bff0ec-481e-466a-843f-68ba5619e6d8/restoring-american-power-7.pdf.
McKinsey & Company. *"And the Winner Is . . .": Philanthropists and Governments Make Prizes Count*. N.p.: McKinsey, 2009.
Meijer, Hugo. *Trading with the Enemy: The Making of U.S. Export Control Policy toward the People's Republic of China*. New York: Oxford University Press, 2016.

Mitchell, R., B. Agle, and D. Wood. "Toward a Theory of Stakeholder Identification and Salience: Defining the Principle of Who and What Really Counts." *Academy of Management Review* 22 (1997): 853–86.

Morrison Foerster. "The Government's Patent Policy: The Bayh-Dole Act and 'Authorization and Consent.'" October 6, 2002. Accessed January 15, 2018. https://www.mofo.com/resources/news/the-governments-patent-policy-the-bayh-dole-act-and-authorization-and-consent.pdf?#zoom=100.

Moynihan, Daniel Patrick, and Larry Combest. "Report of the Commission on Protecting and Reducing Government Secrecy (S. Doc. 105-2)." March 3, 1997. Accessed March 2015. http://www.gpo.gov/fdsys/pkg/GPO-CDOC-105sdoc2/content-detail.html.

Murray, Randy. "Intellectual Property and Technical Data Rights: 'It's About the Money.'" Philadelphia: U.S. Army War College. 2012. www.dtic.mil/get-tr-doc/pdf?AD=ADA593245.

"NASA Budgets: U.S. Spending on Space Travel since 1958." *The Guardian.* February 1, 2010. Accessed 2015. http://www.theguardian.com/news/datablog/2010/feb/01/nasa-budgets-us-spending-space-travel.

NASA. "Benefits from Apollo: Giant Leaps in Technology." July 2004. Accessed 2014. https://www.nasa.gov/sites/default/files/80660main_ApolloFS.pdf.

NASA Technology Transfer Program. "NASA Spinoff." N.d. Accessed May 2018. https://spinoff.nasa.gov.

National Cancer Institute. "Metrics for NCI SBIR Program." National Cancer Institute. 2004. http://dpcpsi.nih.gov/sites/default/files/opep/document/Final_Report_(07-5102-NCI)_10-15-10%20(Year%20Not%20Available).pdf.

National Research Council. *Rising to the Challenge: U.S. Innovation Policy for the Global Economy.* Washington DC: National Academies Press, 2012.

National Science Foundation. "Small Business Innovation Research Program Phase I Solicitation FY-2014 (SBIR) NSF 13-546." National Science Foundation. 2013. http://www.nsf.gov/pubs/2013/nsf13546/nsf13546.htm.

Naval History and Heritage Command. "Aylwin (Destroyer No. 47) II." U.S. Navy. February 2, 2004. Accessed January 4, 2018. https://www.history.navy.mil/research/histories/ship-histories/danfs/a/aylwin-ii.html.

NESTA. "Measuring Innovation." NESTA. http://www.nesta.org.uk/sites/default/files/measuring_innovation.pdf.

Nickerson, Robert C., Upkar Varshney, and Jan Muntermann. "A Method for Taxonomy Development and Its Application in Information Systems." *European Journal of Information Systems* 22 (2013): 335–59.

Obama, B. "Changes Made in Obama Executive Order 13526 on Classified National Security Information." Federation of American Scientists. 2009. Accessed 2014. http://fas.org/sgp/obama/eo13526inout.html.

Office of Management and Budget. "Fiscal Year 2016 Budget of the U.S. Government." 2015. Accessed 2015. http://www.whitehouse.gov/sites/default/files/omb/budget/fy2016/assets/budget.pdf.

Office of Naval Research. "ONR Recruits Defense Community for Online Wargame." Office of Naval Research Corporate Strategic Communications. 2011. http://www.onr.navy.mil/Media-Center/Press-Releases/2011/MMOWGLI-Online-Wargame.aspx.

Office of Science and Technology Policy. "Implementation of Federal Prize Authority: Fiscal Year 2013 Progress Report." Washington DC. May 2014. Accessed March 1, 2018. https://www.challenge.gov/toolkit/files/2017/01/FY2013-Implementation-Federal-Prize-Authority-Report.pdf.

———. "Implementation of Federal Prize Authority: Fiscal Year 2016 Progress Report." Washington DC. July 2017. Accessed March 1, 2018. https://www.challenge.gov/toolkit/files/2017/07/FY2016-Implementation-Federal-Prize-Authority-Report-and-Appendices.pdf.

Oliver, John. "Last Week Tonight with John Oliver: Government Surveillance (HBO)." April 5, 2015. Accessed May 2015. https://www.youtube.com/watch?v=XEVlyP4_11M.

Oman, S., I. Turner, K. Wood, and C. Seepersad. "A Comparison of Creativity and Innovation Metrics and Sample Validation through in-class Design Projects." *Research in Engineering* 24, no. 1 (2013): 65–92.

O'Rourke, Chris. "Mobile Ad Hoc Networking Revamps Military Communications." *COTS Journal*. November 2011. Accessed March 2015. http://www.cotsjournalonline.com/articles/view/102158.

Orszag, P. "Improving Government Acquisition." Executive Office of the President, Office of Management and Budget. 2009. http://www.whitehouse.gov/sites/default/files/omb/assets/memoranda_fy2009/m-09-25.pdf.

Oyen, Timothy. "Stare decisis." Cornell University Law School. Last update March 2017. Accessed 2014. https://www.law.cornell.edu/wex/stare_decisis.

Ozguner, U., C. Stiller, and K. Redmill. "Systems for Safety and Autonomous Behavior in Cars: The DARPA Grand Challenge Experience." *Proceedings of the IEEE* 95, no. 2 (2007): 397–412.

Peck, Merton J., and Frederic M. Scherer. 1962. "The Weapons Acquisition Process; An Economic Analysis." Boston MA: Harvard University, School of Business Administration, 1962.

Pellerin, C. "DARPA Robots to Face Final Challenge in California." Department of Defense. 2014. http://www.defense.gov/news/newsarticle.aspx?id=122672.

Personnel and Document Security Division. "Protecting Classified Information." United States Department of Agriculture. N.d. Accessed 2014. http://www.dm.usda.gov/ocpm/Security%20Guide/S1class/Intro.htm.

Philips. "Open Innovation." Philips. 2014. Accessed December 2014. http://www.research.philips.com/open-innovation/.

"Physical Optics Corp." SBIRsource. 2014. Accessed 2014. http://sbirsource.com/sbir/firms/91-physical-optics-corp.

"Physical Optics Corporation." Small Business Association. N.d. Accessed 2014. https://www.sbir.gov/sbirsearch/detail/271267.

Pike, John. "Selected Exectutive Orders on National Security." Federation of American Scientists. February 20, 2015. Accessed 2015. http://fas.org/irp/offdocs/eo/.
Pluskowski, B. "Innovation Metrics—Part 3." Innovation Excellence. September 12, 2010. Accessed December 2014. http://www.innovationexcellence.com/blog/2010/09/12/innovation-metrics-part-3/.
Porges, Seth. "2007 Crunchies: The Winners." TechCrunch. January 18, 2008. Accessed 2014. http://www.techcrunch.com/2008/01/18/2007-crunchies-the-winners/.
Praetorius, D. "Gamers Decode Protein That Stumped Researchers for 15 Years in Just 3 Weeks." *Huffington Post*. 2011. http://www.huffingtonpost.com/2011/09/19/aids-protein-decoded-gamers_n_970113.html.
President's Blue Ribbon Commission on Defense Management. "A Quest for Excellence [AKA Packard Commission Report]." 1986. Accessed 2018. https://assets.documentcloud.org/documents/2695411/Packard-Commission.pdf.
Priest, Dana, and William M. Arkin. "Top Secret America: A Hidden World, Growing beyond Control." *Washington Post*. July 19, 2010. Accessed 2014. http://projects.washingtonpost.com/top-secret-america/articles/.
Program Assessment and Evaluation Division. "U.S. Marine Corps Concepts and Programs 2013." 2013. Accessed August 2014. http://www.hqmc.marines.mil/Portals/142/Docs/USMCCP2013flipbook/USMC%20CP13%20Final.pdf.
Proposal Exponent. "Federal R&D Funding: Quick Agency Profiles." Proposal Exponent. 2012. Accessed 2014. http://www.proposalexponent.com/federalprofiles.html.
Public Citizen. "Analysis of Executive Order 13292." Public Citizen. 2015. Accessed 2015. http://www.bushsecrecy.org/page.cfm?PagesID=31#A07.
Quirky. "Everything You Need to Know about Quirky." N.d. Accessed November 30, 2017. https://quirky.com/faq/.
"Quirky Pivot Power 6 Outlet Flexible Surge Protector Power Strip (White)." Amazon.com. N.d. Accessed August 2014. http://www.amazon.com/Quirky-Pivot-Outlet-Flexible-Protector/dp/B004ZP74UK.
Rajaram, Srinivasa. *Biometrics: Technologies and Global Markets*. Wellesley: BCC Research, 2016.
Raytheon. "Form 10-Q for RAYTHEON CO/." Raytheon. April 23, 2015. Accessed 2015. http://biz.yahoo.com/e/150423/rtn10-q.html.
———. "Raytheon: Electronic Warfare Overview." Raytheon. 2015. Accessed March 2015. http://www.raytheon.com/capabilities/ew/overview/.
Reporters Committee for Freedom of the Press. "Statutory Exemption." Reporters Committee for Freedom of the Press. N.d. Accessed February 18, 2018. http://www.rcfp.org/federal-open-government-guide/exemptions-disclosure-under-foia/3-statutory-exemption.
Rigby, D., and C. Zook. "Open-Market Innovation." *Harvard Business Review* 80, no. 10 (2002): 80–93. http://www.shadstone-sourcing.com/articles/Open_Market_Innovation.pdf.

"Rules of the Supreme Court of the United States." January 12, 2010. Accessed 2014. http://www.supremecourt.gov/ctrules/2010RulesoftheCourt.pdf.

Sack, H. "Amos Rejects Recent Critique of Amphibious Combat Vehicle." *Marine Corps Times*. June 25, 2014. Accessed December 2014. http://webcache.googleusercontent.com/search?q=cache:T_uLbL6460kJ:archive.marinecorpstimes.com/article/20140625/NEWS04/306250073/Amos-rejects-recent-critique-amphibious-combat-vehicle+&cd=1&hl=en&ct=clnk&gl=us.

"Sandisk Ventures." Sandisk. 2015. Accessed 2015. http://www.sandisk.com/about-sandisk/corporate/ventures/.

Sargent, John F., Jr. "Federal Research and Development Funding: FY2013." December 5, 2013. Accessed 2014. http://fas.org/sgp/crs/misc/R42410.pdf.Sargent, John F., Jr. "Federal Research and Development Funding: FY2018." Congressional Research Service, January 25, 2018. Accessed 2018. https://fas.org/sgp/crs/misc/R44888.pdf.

Satellite Industry Association. "2017 SIA State of Satellite Industry Report." 2017. Accessed January 2, 2018. https://www.sia.org/annual-state-of-the-satellite-industry-reports/2017-sia-state-of-satellite-industry-report/.

———. "Satellite Industry Overview." 2004. http://www.sia.org/wp-content/uploads/2010/11/sat101.ppt.

"SBIR/STTR Submission Site FAQs." Small Business Administration. N.d. Accessed 2014. https://sbir.defensebusiness.org/faqs.

Scott, Alan J. "Sputnik—50 Years Later." *Forum on Physics and Society of the American Physical Society* 36, no. 4 (2007). http://www.aps.org/units/fps/newsletters/2007/october/scott.html.

Sensors Directorate, Air Force Research Laboratory. *Electronic Warfare Technology: Security Classification Guide.* N.p.: Air Force Research Laboratory, 2005.

Serbu, J. "DARPA Challenge Unshreds Destroyed Documents." Federal News Network. November 23, 2011. https://federalnewsnetwork.com/all-news/2011/11/darpa-challenge-unshreds-destroyed-documents/slide/1/.

———. "DIA Releases Technology Wish List to Solve Problems More Directly." Federal News Radio. 2013. http://www.federalnewsradio.com/404/3521041/DIA-releases-technology-wish-list-to-solve-problems-more-directly.

Shah, S. 2000. "Sources and Patterns of Innovation in a Consumer Products Field: Innovations in Sporting Equipment." Sloan Working Paper 4105.

Siehndel, Kathryn. "Freedom of Information Act (FOIA) Request No. F-13-00004." October 2012. Accessed April 2015. https://fas.org/sgp/othergov/invention/stats.pdf.

Silberzahn, Philippe. "Conference on Strategic Surprises at the CIA." Silberzahn & Jones. January 10, 2012. Accessed July 2014. http://silberzahnjones.com/2012/01/10/conference-on-strategic-surprises-at-the-cia/.

Simoes-Brown, D., and R. Harwood. "How to Measure Open Innovation Value—Part 1." 15inno. March 15, 2010. Accessed December 2014. http://www.15inno.com/2010/03/15/howtomeasure/.

———. "How to Measure Open Innovation Value—Part 2." 15inno. March 26, 2010. Accessed December 2014. http://www.15inno.com/2010/03/26/howtomeasurepart2/.

Small Business Association. "About SBIR." N.d. Accessed January 28, 2018. https://www.sbir.gov/about/about-sbir.

Smith, K. H. 2005. "Measuring Innovation." PhD diss., Oxford University.

Snow, Shawn. "The AAV Is Not Dead Yet: The Corps Wants New Tracks to Improve Land and Sea Mobility." April 19, 2018. Accessed May 2018. https://www.marinecorpstimes.com/news/your-marine-corps/2018/04/19/the-aav-is-not-dead-yet-the-corps-wants-new-tracks-to-improve-land-and-sea-mobility/.

Social Security. "Hearing Office Locator, Office of Disability Adjudication and Review." May 2015. Accessed May 2015. http://www.ssa.gov/appeals/ho_locator.html.

Spradlin, D. "Innocentive: Where the World Innovates." N.d. Innocentivehttp://sites.nationalacademies.org/cs/groups/pgasite/documents/webpage/pga_065161.pdf.

Srivastava, Tina Prabha. "Innovation Strategy to Sustain a Technological Edge for National Security and Global Leadership." PhD diss., Massachusetts Institute of Technology, 2015.

———. "Integration Means Change." In *Integrating Program Management with Systems Engineering: Processes, Tools and Organizational Systems for Improving Performance*, edited by Eric S. Rebentisch. Hoboken NJ: Wiley, 2017.

———. *Investigation of the Applicability of the MIT Enterprise Architecting Framework to a Multi-tiered Enterprise: Defense Health Agency Case*. Cambridge MA: MIT Sociotechnical Systems Research Center, 2014.

Srivastava, Tina, Jose Arias, and Victor Piper. "Future Combat Systems Case Study for Analysis of System of Systems Approach." Paper delivered at conference. Rome: International Council on System Engineering (INCOSE) Key Reserve Paper, 2012.

Steipp, C. "Funding Cyberspace: The Case for an Air Force Venture Capital Initiative." *Air and Space Power Journal*. Accessed 2013. 119–28. http://www.airpower.maxwell.af.mil/digital/pdf/articles/Jul-Aug-2013/V-Steipp.pdf.

Stork, K. "About MMOWGLI." Office of Naval Research. June 5, 2014. Accessed December 2014. https://portal.mmowgli.nps.edu/game-wiki/-/wiki/PlayerResources/About+MMOWGLI.

Stouck, Jerry. "Right and Wrong Ways to Use Others' Patents." *National Defense Magazine*. June 2008. Accessed 2015. http://www.nationaldefensemagazine.org/archive/2008/June/Pages/Ethics2293.aspx.

Suh, Eun Suk, and Olivier Ladislas de Weck. "Modeling Prize-Based Open Design Challenges: General Framework and FANG-1 Case Study." *Systems Engineering*. 2018. doi:10.1002/sys.21434.

Takahashi, Dean. "Osama bin Laden's Death Reveals the Value of State-of-the-Art Technology." VentureBeat. May 6, 2011. Accessed July 2014. http://venturebeat.com/2011/05/06/osama-bin-ladens-death-reveals-the-value-of-state-of-the-art-technology/.

Task Force on American Innovation. "American Exceptionalism, American Decline? Research, the Knowledge Economy, and the 21st Century Challenge." December 2012. Accessed July 2014. www.innovationtaskforce.org/docs/Benchmarks%20-%202012.pdf.

Terweisch, Christian, and Yi Xu. "Innovation Contests, Open Innovation, and Multiagent Problem Solving." *Management Science* 54, no. 9 (2008): 1529–43.

Trimble, Stephen. "First Foreign Firm Cracks U.S. Defense Industry's 'Big 5.'" FlightGlobal. February 10, 2009. Accessed 2015. http://www.flightglobal.com/blogs/the-dewline/2009/02/first-foreign-firm-cracks-us-d/.

Truman, Harry S. "Executive Order 10290." *Federal Register*. September 27, 1951.

U.S. Army Medical Department, Medical Research and Materiel Command. Army Technology Objectives (ATOS)-Technology Readiness Levels. U.S. Army. December 24, 2009. Accessed 2014. http://mrmc.amedd.army.mil/index.cfm?pageid=researcher_resources.ppae.atostat.

U.S. Computer Emergency Readiness Team (U.S.-CERT). "Russian Government Cyber Activity Targeting Energy and Other Critical Infrastructure Sectors." March 15, 2018. Accessed May 2018. https://www.us-cert.gov/ncas/alerts/TA18-074A.

U.S. Department of Defense. "Contracts." July 3, 2001. Accessed December 2014. http://www.defense.gov/contracts/contract.aspx?contractid=2053.

———. "DODTechipedia." White House. 2008. Accessed 2014. https://www.whitehouse.gov/open/innovations/DodTechipedia.

———. "Emerging Capability and Prototyping." U.S. Government. 2014. Accessed 2014. http://www.acq.osd.mil/ecp/.

———. "Quadrennial Defense Review 2014." March 4, 2014. Accessed 2014. http://www.defense.gov/pubs/2014_Quadrennial_Defense_Review.pdf.

U.S. Department of Defense, Office of Small Business Programs. "Guidebook for Facilitating Small Business Team Arrangements." 2007. Accessed September 2015. http://www.acq.osd.mil/osbp/docs/dod_OSBP_Guidebook_for_Facilitating_Small_Business_Team_Arrangements.pdf.U.S. Department of Justice. "FOIA Guide, 2004 Edition: Exemption 3." July 23, 2014. Accessed 2014. http://www.justice.gov/oip/foia-guide-2004-edition-exemption-3.

U.S. Department of State. "All about Security Clearances." U.S. Department of State. N.d. Accessed 2015. http://www.state.gov/m/ds/clearances/c10978.htm.

U.S. Export Assistance Center. "Dual Use Licenses." U.S. Government. October 4, 2012. Accessed 2014. http://www.export.gov/regulation/eg_main_018229.asp.

U.S. General Services Administration. "Challenge.gov." 2014. Accessed December 2014. https://www.challenge.gov/about/.

———. "Data.gov." 2014. Accessed December 2014. http://www.data.gov/.

———. "GSA's Challenge.gov Earns Harvard Innovation Award." General Services Administration. 2014. http://www.gsa.gov/portal/content/185155.

U.S. Marine Corps. "U.S. Marine Corps Concepts and Programs 2013." 2013. Accessed February 2015. http://www.marines.mil/Portals/59/Publications/U.S.%20Marine%20Corps%20Concepts%20and%20Programs%202013_1.pdf.

U.S. Patent and Trademark Office. "Design Patent Application Guide." U.S. Patent and Trademark Office. August 13, 2012. Accessed September 2014. http://www.uspto.gov/patents-getting-started/patent-basics/types-patent-applications/design-patent-application-guide.

———. "120-Secrecy Orders [r-11.2013]." November 2013. Accessed 2014. http://www.uspto.gov/web/offices/pac/mpep/s120.html.

U.S. Securities and Exchange Commission. "Updated Investor Bulletin: Crowdfunding for Investors." May 10, 2017. Accessed December 12, 2017. https://www.sec.gov/oiea/investor-alerts-bulletins/ib_crowdfunding-.html.

U.S. Small Business Administration. "Legal Business Size Standards." N.d. Accessed 2014. https://www.sba.gov/category/navigation-structure/contracting/contracting-officials/small-business-size-standards.

———. "Office of Capital Access: Resources." N.d. Accessed 2014. https://www.sba.gov/offices/headquarters/oca/resources/6827.

U.S. Supreme Court. "Department of the Navy v. Egan." February 23, 1988. Accessed 2014. https://supreme.justia.com/cases/federal/us/484/518/.

von Hippel, Eric. *Democratizing Innovation*. Cambridge MA: MIT Press, 2005.

Ven, Andrew H. Van de. "Central Problems in the Management of Innovation." *Management Science* 32, no. 5 (1986): 590–607.

Waitz, Ian A. "Perspective Is Everything." 2009. Accessed 2014. http://web.mit.edu/aeroastro/news/magazine/aeroastro6/intro.html.

Webb, Norman L. "Embedded Research in Practice: A Study of Systemic Reform in Milwaukee Public Schools." *ERIC*, 2000.

Webb, T., C. Guo, J. Lewis, and D. Egel. *Venture Capital and Strategic Investment for Developing Government Mission Capabilities*. N.p.: RAND Corporation, 2014.

White House. "The Cabinet." N.d. Accessed 2014. http://www.whitehouse.gov/administration/cabinet.

———. 1995. "Maintaining Military Advantage through Science & Technology Investment." White House. 1995. Accessed July 2014. http://clinton4.nara.gov/WH/EOP/OSTP/nssts/html/chapt2.html.

———. "2015 National Security Strategy." February 2015. Accessed March 2015. https://www.whitehouse.gov/sites/default/files/docs/2015_national_security_strategy.pdf.

White House, Office of the Press Secretary. "President Obama Lays Out Strategy for American Innovation." White House. 2009. http://www.whitehouse.gov/the_press_office/president-obama-lays-out-strategy-for-american-innovation/.

White House, Open Government Initiative. "DODTechipedia." 2008. Accessed December 2014. http://www.whitehouse.gov/open/innovations/DodTechipedia.

Wiedenman, Nathan. *Adaptive Vehicle Make.* N.p.: DARPA, 2013.

Wikipedia. *Public-key Cryptography.* November 28, 2017. https://en.wikipedia.org/wiki/Public-key_cryptography.

Williams, Mary Ellen Coster, and Diane E. Ghrist. "Intellectual Property Suits in the United States Court of Federal Claims." United States Court of Federal Claims. October 4, 2017. Accessed January 2, 2018. https://www.uscfc.uscourts.gov/node/2927.

Wollerton, Megan. "GE's FirstBuild Facility Opens Its Doors." CNet. July 24, 2014. Accessed 2014. http://www.cnet.com/news/ges-first-build-facility-opens-its-doors/.

Wyatt, Earl. 2014. "Rapid Fielding: A Path for Emerging Concept and Capability Prototyping." 2014. Accessed 2015. http://www.defenseinnovationmarketplace.mil/resources/EmergingCapabilityPrototypingBriefingNDIA.pdf.

Wyatt, T. "In Search of Reasonable Compensation: Patent Infringement by Defense Contractors with the Authorization and Consent of the U.S. Government." *Federal Circuit Bar Journal* 20 (2010): 79. http://cfcbar.org/images/upload/file/WyattFederalClaimsPaper-EXCERPT.pdf.

X Prize Foundation. "Story Ideas." X Prize Foundation. 2015. Accessed February 2015. http://ansari.xprize.org/about/media-room/story-ideas.

Young, Stu. "Lead Systems Integrator Role for Government." NAVAIR. October 28, 2010. Accessed March 2015. www.dtic.mil/ndia/2010systemengr/ThursdayTrack6_11022Young.pdf.

Zacharias, Pat. "When Bomb Shelters Were All the Rage." *Detroit News*. N.d. Accessed 2014. http://undergroundbombshelter.com/news/when-bomb-shelters-were-the-rage.htm.

Zuberi, Bilal. "SBIR/STTR Grants Are Great. 'SBIR Shops' Are Not." May 20, 2014. Accessed January 28, 2018. https://www.luxcapital.com/news/sbirsttr-grants-are-great-sbir-shops-are-not/.

TABLE OF AUTHORITIES

Page numbers with an appended *f* or *t* indicate a figure or table.

Cases

Cramp & Sons v. International Curtis Marine Turbine Co.,
246 U.S. 28 (1918) 110–13, 111*f*, 152nn7–8
Department of the Navy v. Egan, 484 U.S. 518, 527 (1988) 151n33
FAA v. Robertson, 422 U.S. 255 (1975) 152n53
Gargoyles, Inc. v. United States, 113 F.3d 1572 (Fed. Circ. 1997) 118, 153n28
Night Vision Corp. v. United States, 469 F 3d, 1369 (Fed. Cir. 2006) 122–23
Night Vision Corp. v. United States, 68 Fed. Cl. 368 (2005) 122–23
Richmond Screw Anchor Co., Inc. v. United States, 275 U.S. 331 (1928) 152n9, 153n25
Robishaw Engineering v. United States, 891 F. Supp. 1134 (E.D. Va. 1995) 118, 153n26
Zoltek Corp. v. United States, 672 F.3d 1309, 1311 (Fed. Cir. 2012) 126, 154nn40–41

Federal Regulations

Armed Services Procurement Regulation (ASPR) § 9-202 121–22
Code of Federal Regulations
22 CFR § 120.3 . . . . 146n21
22 CFR § 120.10 . . . . 151n45
22 CFR § 121 51
37 CFR § 5.1-5.5 . . . . 145n17
48 CFR § 27.201-1 115–16
78 CFR § 48537 . . . . 149n42
Defense Federal Acquisition Regulation Supplement (DFARS)
252.227-7013 124
252.227-7013(e) 126, 154n31
252.227-7014 124
Federal Acquisition Regulations (FAR)
Clause 52.227-1 . . . . 110, 117, 153n19
Clause 52.227-3 . . . . 153n20
Clause 52.227-5 . . . . 153n21
Subpart 6.302 69

United States Code

5 USC Subchapter II 153n49
5 USC § 552. . . . . . 153n49
10 USC § 130. . . . . . 145n18
10 USC § 2320 125
10 USC § 2410p 149n36
15 USC § 278s 147n33
15 USC § 638. . . . . . 154n34
15 USC § 3710c 154n44
15 USC § 3719 . . . . . . 125, 146–47n30, 146n24, 154n39
15 USC § 3724. . . . . . 125, 147n30
18 USC § 793 . . . . . . 145n20
20 USC § 9801 . . . . . . 146n30
28 USC § 1498 110
35 USC § 181-88. . . . . . 145n17, 153n49
35 USC § 183. 109
35 USC § 271. 114
42 USC § 1861. . . . . . 148n7, 149n3
50 USC § 10, ch. 106, 40 STAT. 411. . . . . . 145n16
50 USC § 403g 151n43
50 USC § 3002 . . . . . . 148n10
51 USC § 20144. . . . . . 146n30

INDEX

Page numbers with an appended italic *f* or *t* indicate a figure or table.

AAAV (Advanced Amphibious Assault Vehicle), 68, 69
A&C. *See* authorization and consent
Acquisition Savings Initiative, OMB, 88, 150n19
Administrative Procedure Act, 153n24
Advanced Amphibious Assault Vehicle (AAAV), 68, 69
Advanced Technology and Projects (ATAP) group, Google, 150n7
Advanced Vehicle Make (AVM), 147n33
aerospace systems, 4, 8
AFLCMC (Air Force Life Cycle Management Center), 32
AFMC (Air Force Materiel Command), 32
AFRL (Air Force Research Laboratory), 33, 37
Air Force, U.S. (USAF), 32, 65–66
Air Force Life Cycle Management Center (AFLCMC), 32
Air Force Materiel Command (AFMC), 32
Air Force Research Laboratory (AFRL), 33, 37
Akula, John, 147n1
Aldrin, Buzz, 44
Allen, Paul, 8
al-Qaeda, 1
America Competes Act (2007), 146n30
America Competes Reauthorization Act (2010), 146–47n30
Amphibious Assault Vehicle (AAV), 46, 54, 68, 147n33
Amphibious Combat Vehicle (ACV), 54, 70
Android operating system, 17
Ansari X Prize, 8, 13*t*, 16, 17
Apollo program, 41, 42–45, 54, 133, 141
Apple, 128
Arkin, William M., 33, 86
Armed Services Procurement Regulation (ASPR; 1955), 121–22
Armstrong, Neil, 44
Army, U.S., 25*t*, 71, 114, 118, 149n35
Army Materiel Command, 114
Army Procurement Regulations (1947), 121
ASPR (Armed Services Procurement Regulation; 1955), 121–22
Astra Zeneca, 10
ATAP (Advanced Technology and Projects) group, Google, 150n7
atomic bomb, 80–81
Atomic Energy Act (2011), 91–92
authorization and consent (A&C): fixing flaws of, 126–29; participation challenge and, 138; patent infringement, U.S. government contractor immunity to, 110–20, 153n25; problem of, 109–20, 111–13*f*, 116*f*; "reasonable compensation" royalties, 113–20, 127–28; selective application of, 153n24; workarounds, 120–21, 122*f*
autonomous driving, 5
AVCI (OnPoint Technologies), 25*t*, 27
AVM (Advanced Vehicle Make), 147n33

B2B (business to business), 13
B2C (business to consumer), 13
BAE, 68, 148n23
Big Five government contractors, 62–63, 114, 119, 121

Boeing, 62, 127
Booz Allen Hamilton, 10
Brown, Alison, 149n34
Bush (George H. W.) administration, 93*t*
Bush (George W.) administration, 2, 83, 93–95, 93*t*
business to business (B2B), 13
business to consumer (B2C), 13

CAGE (Commercial and Government Entity) code, 65
Carter, Ashton, 48; *Keeping the Edge*, 4
Carter administration, 92*t*
CD/V (concept development/validation) phase, 69
Central Intelligence Agency (CIA): classification/declassification of information and, 95, 96, 151n43; first tweet, 33, 35*f*; In-Q-Tel (IQT), 25*t*, 27; as key player in secure U.S. government R&D, 31; website career page, 33, 34*f*
Central Intelligence Agency Act (1949), 96, 151n33
CFT (Cyber Fast Track) program, DARPA, 25*t*, 26, 71
Challenge/Contest/Game, 16*f*, 17
Challenge.gov, 24*t*, 26
Chesbrough, Henry, 7, 11–12, 28, 84
China, R&D investments in, 3, 140
CIA. *See* Central Intelligence Agency
Cisco Entrepreneur in Residence (EIR), 13, 15*t*
Clapper, James R., 87
classified information: access to, 38–40; classification process, 35–37; control over what is classified, 90–92, 151n33; presidential reversals in policy regarding, 92–93*t*, 92–96, 96*f*, 102; redactions in "declassified" documents, 96, 98*f*, 99*f*. *See also* secrecy challenge
Cleland, Brian, 144n11
Cleveland Clinic, 10
Clinton administration, 1, 93–95, 93*t*
Coast Guard, U.S., 149n35
Cold War, 41, 42–43, 45, 90–91
Combest, Larry, 94
Commerce Department, 56
Commercial and Government Entity (CAGE) code, 65
concept development/validation (CD/V) phase, 69
"Confidential" classification, 35
Conrad, Abigail, 144n10
constitutional principles as regulatory regime value, 60
contract vehicles, 70–71
co-option of inventions for national security reasons, 108–20, 111–13*f*, 116*f*
crowdfunding, 14, 16–17, 16*f*, 23, 27–28
crowdsourcing, 13*t*, 49, 52, 125, 130, 144n21, 147n30
Curtis, Charles G., 111*f*
Cyber Fast Track (CFT) program, DARPA, 25*t*, 26, 71

DARPA. *See* Defense Advanced Research Projects Agency
DCA (derivative classification authority), 36
DCAA (Defense Contract Audit Agency), 32–33, 61
DCI (director of central intelligence; now director of national intelligence/DNI), 94–95
DCMA (Defense Contract Management Agency), 61
Deepwater program, 149n35
Defense Advanced Research Projects Agency (DARPA): on complexity of systems, 4; Cyber Fast Track (CFT), 25*t*, 26, 71; defense agencies acquiring technology from, 33; Grand Challenge/Urban Challenge, 25*t*, 26; META software development tool chain, 40–41; national security values in mission statement of, 58; Red Balloon Challenge, 26, 130; Robotics Challenge, 25*t*, 26; secrecy and program manager (PM) tenure under, 81–83; Shredder Challenge, 25*t*, 144n21; Sputnik and founding of, 42, 46; time limits on technology development projects, 150n7. *See also* FANG
Defense Contract Audit Agency (DCAA), 32–33, 61

Defense Contract Management Agency (DCMA), 61
Defense Department. *See* Department of Defense
Defense Federal Acquisition Regulation Supplement (DFARS), 60, 62, 64, 123, 124, 125, 139
Defense Intelligence Agency (DIA), 24*t*, 86, 95
Defense Security Service (DSS), 33
Dell Technologies Capital, 14*t*
Department of Defense (DOD): DOD Acquisition Law Advisory Panel, 96–97, 100–101, 108–9, 115, 129; DODTechipedia, 85–86, 87; *Quadrennial Defense Review*, 145n4; regulatory regime for contractors, 63
Department of Justice (DOJ), 101
derivative classification authority (DCA), 36
design variables: FANG program, 18, 47–51, 48*f*; in open innovation, 18–21, 18*f*, 20*f*
DFARS (Defense Federal Acquisition Regulation Supplement), 60, 62, 64, 123, 124, 125, 139
DIA (Defense Intelligence Agency), 24*t*, 86, 95
Digital Manufacturing Design and Innovation Institute (DMDII), 147n33
Directorate of Defense Trade Controls (DDTC), State Department, 33, 51
director of central intelligence (DCI; now director of national intelligence/DNI), 94–95
DMDII (Digital Manufacturing Design and Innovation Institute), 147n33
DNI (director of national intelligence; formerly director of central intelligence/DCI), 94–95
DOD. *See* Department of Defense
DOD Acquisition Law Advisory Panel, 96–97, 100–101, 108–9, 115, 129
DODTechipedia, 85–86, 87
DOJ (Justice Department), 101
DSS (Defense Security Service), 33
dual-use technologies, 51
Dugan, Regina, 47, 150n7
Dun and Bradstreet Data Universal Numbering System (DUNS), 61
economic growth as regulatory regime value, 59
EFV (Expeditionary Fighting Vehicle), 68–69, 70*f*
EIR (Cisco Entrepreneur in Residence), 13, 15*t*
Eisenhower administration, 92*t*
Elcan Optical Technologies, 114, 119
electronic warfare, 4, 32, 33, 37
Eremenko, Paul, 47
Executive Orders: 8381, 90; 10290, 90–91; 12958, 93, 94; 13292, 93; 13526, 35, 58, 83–84, 91, 93–94
Expeditionary Fighting Vehicle (EFV), 68–69, 70*f*
eXperimental Crowd-Derived Combat-Support Vehicle (XC2V), 49

FAA (Federal Aviation Administration), 101
fairness as regulatory regime value, 60
FANG, 45–54; afterlife of, 147n33; cancellation of program, 53–54, 69, 70; as classified open innovation technology program, 25*t*, 26, 45–47, 52, 53*f*; design variables and, 18, 47–51, 48*f*; FANG-1, 47–54, 53*f*, 123–25, 130; FANG-2, 47, 52, 53*f*; FANG-3, 47, 52, 53*f*; government overreach and, 123–25; non-winning participants, rewards for, 130; regulatory regimes and, 67–71, 70*f*; secrecy challenge and, 80, 81; security classification and data sensitivity issues, 51–54; tiered challenges, 52, 53*f*; XC2V compared to, 49
FAR (Federal Acquisition Regulations), 60, 62, 64, 123, 125, 139
Fast Adaptable Next-Generation Ground Vehicle. *See* FANG
Federal Acquisition Regulations (FAR), 60, 62, 64, 123, 125, 139
Federal Aviation Administration (FAA), 101
FFRDCs (Federally Funded Research & Development Centers), 33
Fifth Amendment, 153n25
FISA (Foreign Intelligence Surveillance Act) courts, 128–29
FMC (later BAE), 148n23

FOIA (Freedom of Information Act; 1966), 101
Foldit, 15t, 19, 50
Ford administration, 92t
Foreign Intelligence Surveillance Act (FISA) courts, 128–29
Freedom of Information Act (FOIA; 1966), 101
functional decomposition in open innovation, 21–23, 22f, 27f
Future Combat Systems program, 149n35

G8, 141
Gabriel, Kaigham J., 150n7
Gagarin, Yuri, 43
Galbraith, Brendan, 144n11
gamification, 5
GAO (Government Accountability Office), 46, 57, 87–89, 124, 150n19
GE: FirstBuild, 15t
General Dynamics, 62, 68–70
genomic mapping, 5
Google: Advanced Technology and Projects (ATAP) group, 150n7; Android operating system, 17; Ventures, 14t
Government Accountability Office (GAO), 46, 57, 87–89, 124, 150n19
Government Purpose Rights (GPR), 124
Government Secrecy Reform Act (1999), 92, 151n32
GPR (Government Purpose Rights), 124
Grasso, Valerie, 149n35

Hiroshima and Nagasaki, 80
Homeland Security, Department of, 94
Hoover Dam, 133
Humphreys, Paul, 144n11

IDIQ ("indefinite delivery, indefinite quantity") contracts, 71
incentives for/obstacles to innovation, 103–31; complex relationship between innovators and U.S. government, 103–4, 130; contractors' employees, low invention disclosure by, 104–8, 106f, 107f, 130; co-option of inventions for national security reasons, 108–9; fixing flawed incentives, 126–31; government overreach issues, 104, 123–26; motivations of innovators, addressing, 129–31; non-winning participants, rewards for, 130; proprietary data, government handling of, 121–23. *See also* authorization and consent
"indefinite delivery, indefinite quantity" (IDIQ) contracts, 71
India, R&D investments in, 3
Indiegogo, 14t, 16
Information Security Oversight Office (ISOO), 94, 95
Innocentive, 10–11, 13t, 16, 124–25
innovation. *See* security and technology innovation
innovator Network, 16f
intellectual property (IP) issues, 9, 17, 19, 52, 64. *See also* incentives for/obstacles to innovation
Interagency Security Classification Appeals Panel (ISCAP), 93–96, 96f, 97f
internal R&D (IR&D), 104
International Traffic in Arms Regulations (ITAR), 33, 51–52, 56
Invention Secrecy Act, 100–101, 145n17, 152n49
invention secrecy orders, 97
IP (intellectual property) issues, 9, 17, 19, 52, 64. *See also* incentives for/obstacles to innovation
IR&D (internal R&D), 104
Iraq war, 91
"iron triangle" of systems engineering, 8, 9f
ISCAP (Interagency Security Classification Appeals Panel), 93–96, 96f, 97f
ISOO (Information Security Oversight Office), 94, 95
ITAR (International Traffic in Arms Regulations), 33, 51–52, 56

job creation as regulatory regime value, 59
Johnson administration, 92t
Justice Department (DOJ), 101

Katrina PeopleFinder, 144n11
Kennedy administration, 42–45, 92t
Kickstarter, 14t, 16
Kiva, 13, 15t, 16

Lead System Integrator (LSI) concept, 74–75, 149nn35–36
LEGO Ideas, 13, 14*t*
Linux, 15*t*
Lockheed Martin, 62, 126, 127
LSI (Lead System Integrator) concept, 74–75, 149nn35–36

Manhattan Project, 80, 81
Manufacturing USA initiative, 147n33
Marine Corps, U.S., 45–47, 49, 50, 53–54, 68–70
Massive Multiplayer Online War Game Leveraging the Internet (MMOWGLI), ONA, 25*t*, 26
MAV (Mojave Aerospace Ventures), 8
MCTL (Military Critical Technologies List), 99–100
META software development tool chain, DARPA, 40–41
Military Critical Technologies List (MCTL), 99–100
MIT: Clean Energy Prize, 15*t*; Instrumentation Laboratory (now Charles Stark Draper Laboratory), 44; Red Balloon Challenge, 26, 130
MMOWGLI (Massive Multiplayer Online War Game Leveraging the Internet), ONA, 25*t*, 26
Mojave Aerospace Ventures (MAV), 8
Moynihan, Daniel, 94
Mudge (Peiter Zatko), 26
Munitions List, U.S., 51

Nagasaki and Hiroshima, 80
NASA: Apollo program, 41, 42–45, 54, 133, 141; Prize Authority, 146n30
National Air and Space Agency. *See* NASA
National Declassification Center, 94
National Defense Industrial Association (NDIA): Small Business Division Subcommittee, 66, 149n34
National Research Council, 4–5
National Science Foundation (NSF), 31, 58
National Science Foundation Act (1950), 58, 79–80, 149n3
National Security Act (1947), 58
National Security Agency (NSA), 95
national security and technology. *See* security and technology innovation
National Security Council (NSC), 95
National Security Strategy (2015), 2–3
Navy, U.S., 1, 25*t*, 26, 38, 112, 151n33, 152n8
Navy Seals, 1
NDIA (National Defense Industrial Association): Small Business Division Subcommittee, 66, 149n34
NeedipeDIA, 24*t*, 25, 85–87, 136
Night Vision Corp., 123
9/11, 93
Nixon administration, 92*t*
Northrop Grumman, 62
NSA (National Security Agency), 95
NSC (National Security Council), 95
NSF (National Science Foundation), 31, 58
nuclear energy, 133

Obama administration, 2, 4, 35, 58, 83–84, 85, 91, 93–95, 93*t*
OCA (original classification authority), 36
OCAs (original classification authorities), 84
Office of Management and Budget (OMB), 88, 150n19
Office of Naval Research (ONA), 25*t*, 26
Office of the Secretary of Defense (OSD), 32–33
OMB (Office of Management and Budget), 88, 150n19
omnibus contracts or bundling, 70
ONA (Office of Naval Research), 25*t*, 26
OnPoint Technologies (AVCI), 25*t*, 27
open competition: secrecy challenge conflicting with, 81
Open Government Initiative, 85–86
open innovation, 5, 7–28; communication and collaboration, 9–10; competition, fostering, 8; degrees of openness, 23–24; design variables in, 18–21, 18*f*, 20*f*; functional decomposition in, 21–23, 22*f*, 27*f*; limitations of, 11–12; participation incentives, 10–11; purposive flow, establishing, 7–8; taxonomy of, 12–23, 13–15*t*, 16*f*, 18*f*, 20*f*, 22*f*; U.S. government forays into, 24–25*t*, 24–28, 27*f*; values associated with, 56–57

operations security (OPSEC), 81
original classification authorities (OCAs), 84
original classification authority (OCA), 36
Osama bin Laden, 1

participation challenge, 59, 134, 138, 139
Patent and Trademark Office (PTO), 37–38
patent infringement, U.S. government contractor immunity to, 110–20, 153n25
patent secrecy orders, 37–38, 39*f*, 40*f*, 96–101, 109
Peer-to-Patent, 144n11
PM (program manager) tenure under DARPA, 81–83
Priest, Dana, 33, 86
prime contractor requirements, 71–74, 73*f*, 103, 137–38, 149n33
private expense, inventions categorized as developed at, 121, 122*f*
prize authority concept, 48, 53, 130, 144n20, 146–47n30
Product Platforming, 16*f*, 17
program manager (PM) tenure under DARPA, 81–83
proprietary data, government handling of, 121–23
PTO (Patent and Trademark Office), 37–38
Public Citizen (organization), 93

Quadrennial Defense Review (DOD), 145n4
Quinn, Barry, 144n11
Quirky, 9–10, 13*t*, 16

Raytheon, 62, 114, 119
Reagan administration, 92*t*
Red Balloon Challenge, DARPA, 26, 130
Red Planet Capital, 25*t*, 27
regulatory regimes, 55–77; bureaucracy, navigating, 61–64, 136–37; competing values and, 57–60; co-option of inventions for national security reasons, 108; entrenched policies and relationships in, 67–71, 70*f*; established prime contractor requirements, 71–74, 73*f*, 103, 137–38, 149n33; importance of changing, 134–36; for large contractors, 62–63; Lead System Integrator (LSI) concept, 74–75, 149nn35–36; participation challenge and, 59, 134, 138, 139; policy tensions and, 55–57; SBIR shops, 75–77, 76*f*, 77*f*, 137–38; secrecy challenge and, 59, 83–88, 134, 138–39; for small businesses and start-ups, 61–62, 64–67, 71–75, 73*f*, 149nn33–35
requests for proposals (RFPS), 86
Revitalize American Manufacturing and Innovation Act (2014), 147n33
RFPS (requests for proposals), 86
Robotics Challenge, DARPA, 25*t*, 26
Roche Diagnostics, 10–11
Roosevelt (Franklin D.) administration, 90
Roosevelt, Franklin D. (as secretary of the navy), 112, 152n9
rule of law as regulatory regime value, 60
Russia: use of technology by, 140. *See also* Soviet Union
Rutan, Burt, 8

SAM (System for Award Management), 61
Samsung, 128
Sandisk Ventures, 13, 14*t*
SAPS (Special Access Programs), 35–36, 86–87
satellite technology, 42, 56–57, 57*f*
SBA (Small Business Administration), 32, 61
SBIR (Small Business Innovation Research), 32, 33, 71, 123
SBIR shops, 75–77, 76*f*, 77*f*, 137–38
SCGS (security classification guides), 36, 37
Schubert, William, 119
SCI (Sensitive Compartmented Information), 36
SEC (Securities and Exchange Commission), 17
secrecy challenge, 79–102; control over what is classified, 90–92, 151n33; DARPA program manager (PM) tenure and, 81–83; as endemic constraint, 138; exceptions to limiting competition and, 87–90; open competition and, 81; overcoming, 101–2, 138–39; patent secrecy orders, 96–101; presidential reversals in policy and, 92–93*t*, 92–96, 96*f*, 102; regulatory regimes and,

59, 83–88, 134, 138–39; relationship of national security and secrecy, 79–81; selective declassification to leverage open technology innovation, 84–85, 138–39; silo effect and, 81, 85, 131
"Secret" classification, 35
secure U.S. government R&D, 5, 29–40; access to classified information, 38–40; basic research, applied research, and development, 29; budget and scale of, 29, 31; classification of information, 35–37; defined and described, 33–35; key players in, 31–33, *32f*, 145n8; secrecy orders, 37–38, *39f*, *40f*; technology readiness level (TRL) and, 29–31, *30f*. *See also* regulatory regimes; successes and failures; *specific programs*
Securities and Exchange Commission (SEC), 17
security and technology innovation, 1–6, 133–41; American leadership in, 1–3, 133–34, 140–41; endemic constraints, overcoming, 136–38; incentives for/obstacles to, 103–31; national security as primary value, 57–59, 79; open innovation, 5, 7–28; regulatory regimes for, 55–77; secrecy and, 79–102; secure innovation by U.S. government, 5, 29–40; strategies for, 1–6, *4f*; successes and failures of, 41–54; in unstable geopolitical scene, 133, 140. *See also* incentives for/obstacles to innovation; open innovation; regulatory regimes; secrecy challenge; secure U.S. government R&D; successes and failures
security classification guides (SCGs), 36, 37
security classification in U.S., 35–37
security clearances, 38–39, 146n23
Sensitive Compartmented Information (SCI), 36
September 11, 2001, 93
Shredder Challenge, DARPA, *25t*, 144n21
Silicon Valley, 40
silo effect, 81, 85, 131
single entity/single contractor dependence, 149n35
Small Business Administration (SBA), 32, 61
Small Business Innovation Research (SBIR), 32, 33, 71, 123. *See also* SBIR shops
Small Business Research and Development Enhancement Act (1992), 145n9
smartphones, 7, 128
Soviet Union: Cold War, 41, 42–43, 45, 90–91; fall of, 3; Space Race with, 42, 43, 46. *See also* Russia
Space Race, 42–45
Special Access Programs (SAPS), 35–36, 86–87
Sputnik, 42, 43, 46
State Department, 33, 51, 56
Stevenson-Wydler Technology Innovation Act (1980), 146n30
successes and failures, 41–54; FANG, 45–54; NASA Apollo program, 41, 42–45, 54
Supreme Court, U.S.: on authorization and consent (A&C), 110, 112; *Egan* decision on classification authority, 151n33; on FOIA, 101
System for Award Management (SAM), 61
systems engineering, 8, *9f*

Taft, William H. (as Supreme Court justice), 110, 115, 153n25
takings and patent infringement, 153n25
Task Force on Innovation, 3
technology and national security. *See* security and technology innovation
technology readiness levels (TRLS), 29–31, *30f*
telecommunications satellite technology, 56–57, *57f*
Tenebraex, 114–15, 118, 119
theoretical IP transfer challenges, 124–25
Threadless, 13, *14t*
Topcoder, *14t*, 19
"Top Secret" classification, 35, 79
trade secrets, designating innovations as, 120
Trading with the Enemy Act (1917), 145n16
TRLS (technology readiness levels), 29–31, *30f*
Truman administration, 90–91, *92t*
Trump administration, 31

United Defense (UDI; later BAE), 148n23
United Nations, 141

USAF (U.S. Air Force), 32, 65–66
U.S. departments, agencies, and organizations. *See specific entity, e.g.* Air Force, U.S., Commerce Department
User-Centered Innovation, 16*f*, 17
U.S. government: open innovation, forays into, 24–25*t*, 24–28, 27*f*. *See also* secure U.S. government R&D
USS *Aylwin*, 112*f*
USS *Warrington Destroyer*, 113*f*

Venture Capital (VC) Arm, 16*f*

Washington Post, 33–35, 86–87
Wiedenman, Nathan, 4, 47
WikiLeaks, 91
World War I, 37, 110
World War II, 45–46, 90

XC2V (eXperimental Crowd-Derived Combat-Support Vehicle), 49
X Prize Foundation, 8

Zatko, Peiter (Mudge), 26
Zazzle, 15*t*